group of architects 大象
設計

大象设计 2018–2023

goa大象设计 编著

同济大学出版社·上海

序

阅读 goa 大象设计

李翔宁
同济大学建筑与城市规划学院院长、教授
2023年8月

在中国当代建筑设计的行业版图之中，国有建筑设计院、独立建筑事务所与大型民营设计院代表了三类重要的组织模式。如果说国有建筑设计院以雄厚的技术力量和资源承担了大规模的工程项目，独立建筑事务所在个人化实践风格的主导下寻求独特定位，那么大型民营设计院则以多元性的集体创作回应不同的市场需求与项目类型。作为民营设计院的翘楚，goa 大象设计呈现了这样一种发展轨迹：在国有设计单位接受了职业初期的价值观与工作方式塑造，而后选择投身于市场化的行业浪潮，并逐渐成长为富有影响力的建筑团队。理解这种由"单位"向"公司"的跨越，或许构成了阅读 goa 大象设计的起点。正如"Group of Architects"这个名字所暗示的，团队合作的组织模式构成了公司运作的基础，既避免了建筑师陷入过度个人风格化的陷阱，也在集体与个人之间找到了某种自洽的平衡。

但是，这种跨越不仅体现在体制与模式的转变上，更重要的意义是获得一种综合性的设计视野。以专业的组织制度与建筑技艺为核心，goa 大象设计面向不同的市场需要、人群画像与环境特征发展出连接的可能性，展现出一种多元影响要素交融的实践策略。在二十余年的创作历程中，goa 大象设计以多重身份介入了多个项目的全生命周期：参与房地产策划，在设计环节整体统筹建筑、结构、景观与室内专业，作为业主参与市场投资与运营，并在一些项目中自身成为建筑的使用者。这些经验使 goa 大象设计跳出了房地产商、建筑师、业主或使用者的单方视角，将综合产业链的需求融入设计之中，进而以整体性的框架思考建筑在市场中的合适定位与发展方向。

基于对社会与市场需求的敏锐观察，goa 大象设计注重建筑的服务性质，满足市场需求与文化表达，使建筑具有高度的实用性与市场竞争力。这样的策略使人联想起贝聿铭的建筑生涯——同样在职业初期与房地产商合作，相较于秉承纯粹现代主义观念的建筑师，贝聿铭并不一味追随以空间形式为核心的建筑自治性叙事，其对人性化需求的关注使其赢得了许多重要的设计委托。与此类似，goa 大象设计并不持有市场决定论的唯一预设，而是始终将如何响应"人"的需求放置在设计实践的核心。例如，蓝城陶然里项目在实现跨代际混居的适老型社区的理念上做出了积极的尝试与创新，华润亚奥城则通过打造小街密路规划结构之下以步行街为纽带的社区，增进了居民之间的交流与互动。

在不同需求相互角力与竞争的现实语境中，goa 大象设计在建筑学科的基本立场与对社会的回应之间架起桥梁，并在更广泛的层面上与城市、社会与文化互动，使其成为了一种如萨拉·怀汀所言的"投射式实践"。这一定位为他们的实践赋予了都市性的维度，并在其作品中实现了双重意义上的清晰表达：在空间层面，充分比选建筑与城市形态关系，通过空间节点、界面与场域的推敲实现建筑组群的精心塑造；在功能层面，从功能混合与公共性的角度审视建筑之于城市发展的意义，使建筑成为承载多样化的人群活动的微型城市。

天目里综合艺术园区的落成历程也折射出 goa 大象设计的都市性思考。该项目由意大利建筑师伦佐·皮亚诺担纲设计，goa 大象设计以执行建筑师的身份为项目提供全过程的技术推动，同时以投资方之一的身份参与项目的策划定位决策。落成后的园区如同一座多元融合的城市客厅。建筑不再是孤立的形式，而是与社会、文化与艺术相互交织的动态体现。天目里建成开放以来，迅速成为集聚公共活力的城市目的地。

同时，goa 大象设计的实践体现了建筑师对于在地性的敏感体验与表达。以杭州为中心，他们将江南文化融入建筑设计，结合西方的几何秩序与东方的简约美学，以极高的完成度呈现出对当地文化的尊重与理解。在木守西溪度假酒店中，建筑师创造性地塑造建筑与水岸界面的丰富变化，并使用玻璃幕墙与廊下空间模糊室内与室外的边界，获得了建筑与自然景观之间的尺度适宜性。

Preface

Reflecting on the Architectural Journey of GOA

Li Xiangning
Dean and Professor, College of Architecture and Urban Planning, Tongji University
August 2023

In the landscape of contemporary Chinese architectural design, three primary organizational models have emerged as dominant: state-owned architectural design institutes, independent architectural firms, and large private design institutions. State-owned design institutes manage massive construction projects, leveraging their significant technical expertise and resources. Independent firms differentiate themselves by developing individualized design philosophies tailored to their specific niche. In contrast, large private design institutions, like GOA, offer diverse collaborative creative endeavors, responding to different market needs and project classifications. GOA's trajectory is notable. Having initially shaped their professional values within state-run institutions, they later embraced market-driven design currents to evolve into an influential architectural entity. Understanding the transition from the "danwei" (work unit) system to the "corporation" structure is crucial in comprehending the initial phase of GOA's journey. The firm, known as Group of Architects, places a strong emphasis on collaborative teamwork, as implied by its name. This approach achieves a harmonious equilibrium between the collective and the individual, which would otherwise be an overemphasis on stylistic individualism.

However, this transition isn't just evident in shifts in systems and models but more importantly, signifies the acquisition of an integrated design perspective. Anchored in professional organizational structures and architectural approaches, GOA has continually sought and identified opportunities tailored to diverse market needs, user demographics, and environmental contexts. This strategy involves the convergence of multiple influential elements. For over two decades, GOA has engaged in the complete life cycle of numerous projects, assuming various roles including real estate planning, comprehensive coordination of architectural structures, landscape, and interior design in addition to investments and operations in the industry, and end-users of projects. Such experiences have enabled GOA to transcend limited perspectives while synchronizing its designs with broader requirements within the industry, thereby comprehensively establishing an appropriate market positioning and developmental path in the field of architecture.

GOA's keen sensitivity to societal and market demands underscores their emphasis on the functional essence of architecture, marrying practical utility with cultural expression and market relevance. Such a strategy is reminiscent of I. M. Pei's career as an architect, who also collaborated with real estate developers early in his career. Pei's attention to addressing humanistic needs, in contrast to the approach of purely modernist architects who prioritized a narrative of architectural autonomy centered on formal space, resulted in him receiving substantial design commissions. Similarly, rather than adhering rigidly to the principles of market determinism, the philosophy of GOA prioritizes human needs as the fundamental basis for design. For instance, the Bluetown The Kidult exemplifies a proactive effort to establish an age-friendly community conducive to intergenerational cohabitation, while the Hangzhou Asian Games Technical Official Village nurtures connection and interaction by incorporating a sophisticated pedestrian texture.

Navigating the complexities of conflicting values, GOA bridges fundamental architectural tenets with societal responsiveness through deep interactions with the urban, societal, and cultural fabric, exemplifying what Sarah Whiting describes as a "projective" architecture practice. This role imbues GOA's work with urban characteristics while facilitating a dual-layered clarity of design expression. Spatially, the evaluation and juxtaposition of architecture in relation to urban form involves a meticulous consideration of spatial elements for sculpting architectural complexes, such as nodes, interfaces, and domains. Functionally, the programmatic composition and degree of public accessibility establish the architectural significance in urban development, transforming the built environment into a microcosm of the larger urban context and providing space for diverse human activities.

伴随着近年来城市化进程逐渐步入存量更新时代，goa大象设计的实践触角也深入传统风貌区域的城市更新。在祥符桥传统风貌街区、里直街等项目中，goa大象设计的方式是尝试在传统与现代的风格之间寻找适宜的变奏与平衡，其关键在于既保留历史街区的丰富建筑遗产与共同记忆，又能使其与现代社会的需求相适应。而在仁恒海上源项目中，则会看到goa大象设计着意将杨树浦港的工业棕地的整合性再生纳入住宅地块的城市界面与公共空间设计，充分体现建筑师对城市人居环境的价值认知与人文关切。

　　纵观全球建筑行业中具有典范地位的设计公司，其中既包括AECOM这样涵盖工程总包施工与建设的综合性全球企业，也有以KPF、SOM、Gensler为代表的高度专业化的大型建筑事务所。尽管中国与西方的建筑设计公司在组织模式与体制上存在显著差异，但若将以上公司视为中国大型国有建筑设计院与民营建筑企业的共同标杆，或许并非言过其实。作为中国民营建筑企业的重要代表之一，goa大象设计深刻理解市场需求与人们对建筑服务的期望，其实践作品充分展示了建筑设计、社会需求与文化内涵之间的紧密关联，使其不仅在设计构思上呈现出独特的创造性，还展现了对文化因素的敏感性与设计施工的高完成度，同时兼顾业主投资与运营的综合性考量。正是goa大象设计的多元性给出了如何平衡设计品质与社会价值的有效范本，展示了其与全球顶尖建筑事务所相比肩的潜力，并在开放的边界中，为未来的建筑实践提供了可持续性的思考空间与发展愿景。

GOA's urban vision is demonstrated through the establishment of the OōEli Comprehensive Art Park. Led by RPBW (Renzo Piano Building Workshop), this project fulfilled both GOA's role as executive architects, responsible for technical supervision, and as investors engaged in project positioning. OōEli emerged as a versatile urban parlor, more than merely an isolated form; the architecture is intricately interwoven with the dynamic of society, culture, and art. It has become an urban magnet shortly after its inauguration.

In another direction, GOA's work demonstrates the architects' perceptive comprehension and delicate depiction of local context. Based in Hangzhou, GOA weaves Jiangnan culture into its architectural designs, reflecting a profound reverence for local heritage through a fusion of Eastern minimalist aesthetics and Western geometric principles. In Muh Shoou Xixi, the utilization of glass facades and continuous porch spaces shape the interfaces between the building and water, resulting in a balanced and cohesive relationship between architecture and nature by erasing indoor-outdoor boundaries.

As urbanization in recent years shifts towards rejuvenation, GOA has directed its attention towards revitalizing historic districts. In projects such as the Xiangfu Bridge Historic District Renewal and Lizhi Street, GOA negotiates the interplay between tradition and modernity by preserving valuable the area's architectural heritage and collective memories while adapting them to the demands of contemporary society. Within Yanlord Arcadia, GOA seeks to achieve an integrated renewal of Yangshupu Port's industrial brownfield into the design of the residential plot's urban interface and public spaces, reflecting the architect's profound respect for the urban living environment on top of their humanistic approach to design.

Surveying the architecture industry on a global scale, it's easy to identify certain design firms that stand out as paragons. This includes international entities, like AECOM, that manage the comprehensive engineering and construction process, as well as highly specialized architectural giants represented by KPF, SOM, and Gensler. While there are evident disparities between the organizational structures of Chinese and Western architectural firms, viewing the aforementioned companies as benchmarks for both China's major state-owned architectural design institutes and private enterprises cannot be overstated. GOA, as a pivotal representative of China's private architectural entities, possesses a comprehensive understanding of market demands and expectations associated with architectural services. GOA's portfolio vividly highlights the symbiotic relationship between architectural design, societal needs, and cultural significance. It exhibits not only a distinctive and inventive flair in concepts but also a nuanced sensitivity to cultural influences and superior craftsmanship in design and construction while considering stakeholder investment and operational aspects in a comprehensive manner. GOA's multifaceted approach illustrates the integration of design quality with societal value, positioning it as a potential contender among the elite architectural firms worldwide and establishing a sustainable vision for future architectural endeavors within open boundaries.

目录 Content

010　出发点——陆皓 Inception——LU Hao
018　访谈 goa 大象设计：在变化中拥抱设计的价值 Interview with GOA: Embracing the Value of Design in Changes
024　goa 大象设计总部 GOA Headquarters

混合使用开发 For Mixed-use

032　上海西岸金融城 G 地块 Shanghai West Bund Financial City - Plot G
036　鹿山时代 Lushan Times
040　深圳万创云汇 01—03 地块 Shenzhen Vanke Cloud Gradus - Plot 01-03
044　上海长风中心 Shanghai Changfeng Center
048　重庆启元 Century Land Chongqing
052　恒力环企中心 Hengli Global Enterprise Center
054　杭州西动所上盖及周边区域综合开发 Hangzhou West High-speed Train Maintenance Base Superstructure
056　望江中心 TOD Wangjiang Center TOD
060　滨耀城 Colorful City
064　天目里 OōEli

办公 & 产业 For Working

076　浙商银行总部 China Zheshang Bank Headquarters
082　宇视科技总部 Uniview Headquarters
088　西子智慧产业园 Xizi Wisdom Industrial Park
094　杭州东站花园国际 Hangzhou East Railway Station Garden International
098　海口五源河创新产业中心 Haikou Wuyuanhe Intelligent Creative Collective
100　石家庄中央商务区办公楼 Shijiazhuang CBD Office Towers
102　宁波智造港芯创园 Ningbo Intelligent Manufacturing Port
106　江南布衣仓储园区 JNBY Warehousing Logistics Park
108　台州数字科技园 Taizhou Digital Technology Park

城市更新 For Urban Renovation

114　祥符桥传统风貌街区 Xiangfu Bridge Historic District Renewal
122　上海北外滩 32 街坊更新 Shanghai North Bund Neighborhood 32 Renewal
126　弘安里 Hong'anli
128　中海顺昌玖里 China Overseas Arbour
130　里直街 Lizhi Street

居住 For Living

- 136　绿城湖境云庐 Greentown Hangzhou Oriental Villa
- 142　仁恒海上源 Yanlord Arcadia
- 146　华润亚奥城 CR Land the Century City
- 150　绿地海珀外滩 Greenland Hysun Bund
- 152　融创长乐雅颂 Sunac Changle Yasong
- 156　绿城外滩兰庭 Greentown the Bund Garden
- 160　融创滨江杭源御潮府 Sunac Binjiang Imperial Mansion
- 164　绿城春风金沙 Greentown Hangzhou Lakeside Mansion
- 168　华润武汉瑞府 CR Land Wuhan Park Lane Mansion
- 170　蓝城陶然里 Bluetown the Kidult
- 172　绿城空中院墅 Greentown Sky Villa

度假 & 休闲 For Leisure

- 178　阿丽拉乌镇 Alila Wuzhen
- 186　木守西溪 Muh Shoou Xixi
- 192　湘湖逍遥庄园 Xianghu Xiaoyao Manor
- 198　杭州远洋凯宾斯基酒店 Kempinski Hotel Hangzhou
- 202　苏州狮山悦榕庄 Banyan Tree Suzhou Shishan
- 206　德清莫干山洲际酒店 InterContinental Deqing Moganshan
- 210　湘湖陈家埠酒店 Xianghu Chenjiabu Hotel
- 212　青岛藏马山酒店 Qingdao Cangmashan Hotel
- 214　既下山大同 SUNYATA Hotel Datong

城乡协同 For Suburban-urban Mutualism

- 220　阳羡溪山 Yangxian Landscape
- 224　曲水善湾乡村振兴示范区 Qushui Shanwan Rural Revitalization Demonstration Area
- 228　曹山未来城古桥水镇 Caoshan Future City Guqiao Water Town
- 230　张謇故里小镇柳西半街 Jianli Town Liuxiban Street

公共设施 For Public

- 236　雅达剧院 Yada Theater
- 242　沪杭高速嘉兴服务区 G60 Expressway Jiaxing Service Area
- 246　建德市文化综合体 Jiande Cultural Center
- 250　舟山绿城育华幼儿园 Zhoushan Greentown Yuhua Kindergarten
- 256　飞鸟剧场 Earth Valley Theater
- 260　天台山雪乐园 Tiantaishan Snow Park

- 265　核心团队 Core Team
- 271　项目资料 Project Data
- 283　项目列表 Chronology

出发点

陆皓
goa 大象设计总裁、总建筑师
2023年8月

goa 大象设计成立 25 周年之际，同事们建议我写些东西以留念，想了一下，还是从我们的出发点说起吧。

1998 年春，包括我在内的六位"年轻人"，从杭州市建筑设计院办理了离职。创业之初我们得到了绿城集团宋卫平董事长的支持和信任，主要为绿城集团做设计。在此之前我们完全没有住宅设计的经验和认识，现在回忆起来也算是无知者无畏了。

杭州桂花城是我们的第一个大型住宅社区项目。接到设计任务时，国内的房产品刚刚起步，基本没有经验可循，我们便翻阅了当时所能接触到的国外住宅设计资料。恰逢 20 世纪八九十年代新城市主义在西方大行其道，我们便结合国内住宅的户型要求，以国外成功的新城市主义项目为蓝本，演化出了桂花城系列。

今天回头去看这个项目，中心轴线上的一系列公共空间——包括大草坪和广场、会所和幼儿园、由商业空间构成的社区中心——历经二十余年依然发挥着显著的作用。那些在阳光与绿色环抱之中嬉戏休憩的人们，给予了作为建筑师的我们至深的满足和骄傲。

Inception

LU Hao
President and Principal of GOA
August 2023

For GOA's 25th anniversary, my colleagues suggested I write something to commemorate the journey we have taken to arrive here. To this end, allow me to bring you back to our humble beginnings.

In the spring of 1998, six young individuals, myself included, departed from the Hangzhou Architectural Design and Research Institute to embark on a new venture. Initially, Mr. Song Weiping, the chairman of Greentown China, supported and entrusted us to design buildings for the company. Despite lacking any prior experience in or knowledge of residential design, we leaned into the opportunity with courage and determination. It was a clear case of youthful naivety leading the charge forward.

Our inaugural large-scale residential community endeavor was Hangzhou Osmanthus Town. China's real estate industry was still in its relative infancy at this point, leaving us without an established precedent or even clear guidance. But we pressed on. We studied all available materials on foreign residential design that we could get our hands on. During the 1980s and 1990s of the 20th century, the New Urbanism movement had been gaining significant traction in Western countries for some time. Consequently, the Osmanthus Town Series was primarily developed by aligning domestic housing demands with successful overseas New Urbanism paradigms.

可能连我们自己也不曾想到，杭州桂花城的设计经历会成为一颗影响深远的种子。从初创时的"六人组"到如今的大象设计，我们这群建筑师在人居设计领域长期耕耘，越发清晰的是这样一个核心认识：建筑的价值不仅仅在于建筑学层面的空间艺术和美学造诣，更在于其为人的生活带来的切实影响，建筑的使命在于为生活其中的人带来美好生活的可能性。

围绕这一核心认识，我们实践的出发点有几个层面。

自然环境的打造是最早被体认到的。大象设计的早期项目常常以住区中心自然环境的打造为最重要的着力点。我们总是努力将设计与用地的气候、地形、植物、水域等因素密切结合起来。建筑和自然的紧密关系是这些早期实践的一个显著标志。

建筑的在地性及文化认同则是另一个重要命题，自创业之初便伴随我们左右。2003年西子湖四季酒店的设计中，我们通过一系列研究将江南园林的意蕴转译于现代生活场景中，自此之后，团队不断在类型各异的实践中探索传统文化的当代表达。在我们的理解中，一个"在地"的建筑师需要做到对投资水平、技术能力的主动适应和熟练应用，而重视"文化认同"则意味着主动寻求与建筑使用者、社会大众之间的价值共鸣，这两点都是我们十分关注的。

Revisiting this project today, it's easy to still witness this initial design pattern with a central axis connecting public spaces: the grand lawn and plazas, the recreation club and kindergarten, and the integrated commercial community center. Even after two decades, these public spaces still provide immense value, offering solace, sunshine, and verdure to the community which inhabits and supports them. As publicly-minded architects, it is impossible for us to witness these vibrant scenes and not be filled with profound joy and pride.

Little did we imagine that the design of Hangzhou Osmanthus Town would sow such a profound seed. From the embryonic days of "Squad Six" to today's GOA, our group of architects has remained dedicated to enhancing human habitats. It has become increasingly evident to us that architectural value transcends spatial strategies and aesthetic attributes; it resides in tangible contributions to people's lives. The crux of architectural design lies in enabling living possibilities.

Anchored in this fundamental understanding, our design philosophy has been manifested through several key initiatives.

Our earliest priority was to establish the natural environment as the heart of residential zones. We aimed to integrate natural conditions into the design, encompassing climate, topography, water features, vegetation, and other factors. As a result, the intimate interplay between architecture and nature is significantly evident in GOA's early works.

随着实践的累积，包括我自己在内的诸多团队成员得以获得一种有趣的体验——成为自己笔下住宅、酒店等项目的使用者。我们跳脱出职业训练的一般性框架，建立起真实的使用者视角。这样的亲身经历不仅帮助我们更为成熟地应对建筑使用方面的效率和功能问题，更为我们的创作打开了一个新的思考维度——环境如何更好地激发使用者的积极活动，尤其是高质量的公共活动。

我们提出的"鲜活社区"理念正是这一视角的具体表达。基于这一理念，我们以住区为切入点，主动为目前国内城市生活的贫乏和城市空间的割裂寻找建筑层面的解决办法：在保留住区既有居住习惯和管理模式的同时，打通住区和城市之间的联系，让住区的公共生活融入整个城市中。这些探索性尝试并非一朝之思，而是源于若干年来我们长期代入的"使用者"立场。

近年来，随着实践的多元化延展，我们将人居类项目中的经验应用至其他各种类型的项目实践中，取得了一些成果；对于"出发点"的理解也由此愈发清晰、深入。总的来说，我们是这样的实践者：无论面对何种建筑类型，总是视建筑空间环境为一个整体；我们关心环境的完整性和序列性，更关心使用者于环境中所获得的与自然、与文化、与他人互动的体验。

From the outset, we have significantly emphasized site specificity and cultural identity within our leading philosophy. This guiding ethos found its highest expression in our 2003 design of the Four Seasons Hotel at West Lake, where we reinterpreted the classical Jiangnan Garden through a modern lens. Subsequently, we embarked on a journey of exploring contemporary translations of traditional cultures across various projects. We have long persisted in valuing "localization" by consistently adapting our vision to the evolving investment and technology landscapes. Similarly, we have sought to honor "cultural identity" by delivering design values that resonate with all of our many stakeholders. Both are our pivotal considerations.

As we take on more projects, we find the exciting opportunity to experience our designs as users ourselves, particularly those of residential and hotel spaces. This shift in perspective has greatly enhanced our design process, allowing us to better address architectural functionality and efficiency. Additionally, this evolving comprehension of the environmental impacts on positive communal activities has propelled us into a new dimension of design contemplation.

The "Living Community" concept is a concrete embodiment of this shift from a professional to an experiential perspective. We initiated this concept with residential communities, expanding to encompass a series of urban solutions within an architectural framework. This approach consistently tackles recurring urban scenarios and fragmented spaces while aiming to preserve the existing lifestyle and operational structure. Simultaneously, we endeavor to forge connections between residential spaces and the expanding urban contexts, thereby weaving public life into the urban fabric. Though complex, this exploratory initiative is far from arbitrary and has continued to evolve from our extensive experience and comprehension as "users".

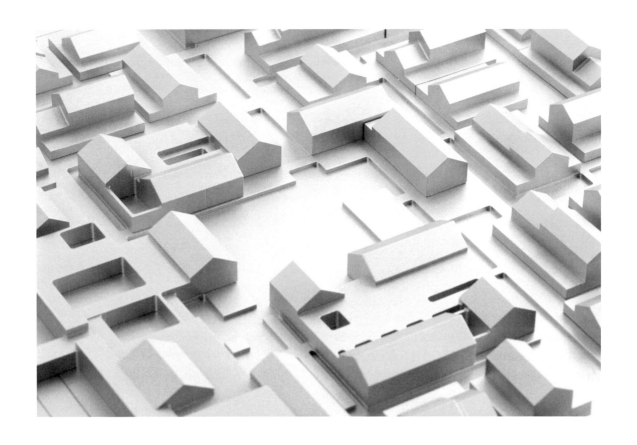

与实践的多元化趋势并行的，是大象设计团队始终如一的多样性特质。这得益于我们在创业之初定下的两项基本制度——贯彻始终的合伙人制度和以合伙人为中心的工作小组制度。制度的背后是两项基本的共识：其一，我们认为建筑设计从出发点到成果之间的思维连续性至关重要，因此设计中的集体合作必须在有能力的决策者的全程掌控之下，这种掌控关系正如同指挥之于交响乐，导演之于电影；其二，我们拥抱不同视角的多样化的设计解决之道。时至今日，公司的组织架构及运行方式皆围绕这两项基本制度展开，它深刻地决定了团队今天的成就和将来的发展。

回头看，这些年 goa 大象设计所取得的成果和经验得益于时代的机遇，得益于团队对设计的努力和专注，也得益于广大业主的信任和支持。未来从过去而来，对于出发点的坚守与探寻会带领我们在时代浪潮中继续前行。

Throughout the years, our involvement in diverse project typologies has facilitated the transfer of our residential expertise to other architectural realms. Along this journey, we have been ever grateful for the many commendations we have received. More importantly, however, has been the deepening understanding of our origins. As practitioners, we envision architectural environments as harmonious entities, emphasizing spatial coherence and discipline and, most notably, the interplay between users and nature, culture, and society.

GOA's consistent commitment to variety and exploration has aligned with the expanding diversity in our work. This alignment can be attributed to two foundational systems established since our inception: partnership and partner-oriented design leadership. Both have been rooted in two simple consensuses from the very beginning. First, we hold that logical consistency is key in architectural design, necessitating sophisticated decision-makers to steer the course, similar to an orchestra conductor or a film director. Second, we embrace diverse design solutions from different perspectives. These two bedrock systems, embedded in GOA's organizational structure and operations, profoundly influence the company's achievements and future development.

Reflecting on our journey, GOA's accomplishments and growth emerge as a synthesis of various elements. We have a lot to be grateful for: We appreciate the opportunities of this thriving era; we appreciate the ongoing trust and support of our clients; and we remain extremely grateful for the unrelenting dedication of our team. As we venture forth, we remain forever faithful to the principles that informed our inception, allowing them to serve as our guiding stars as we navigate the ever-changing tide of time.

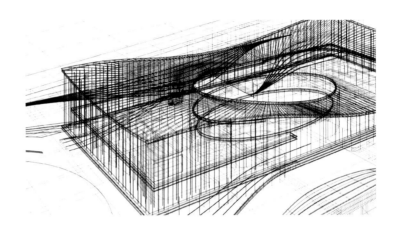
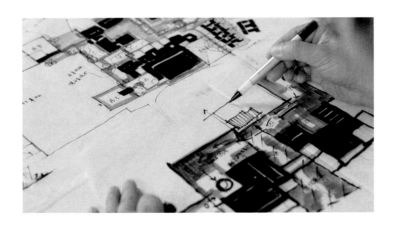

访谈 goa 大象设计：
在变化中拥抱设计的价值

访谈者：
Architecture China 期刊

受访者：
陆皓，goa 大象设计总裁、总建筑师
何兼，goa 大象设计执行总裁、总建筑师
何峻，goa 大象设计执行总裁、总建筑师

1 公司的创立与发展节点

问：从行业角度来说，当时创办公司的动力和目标是什么？

何兼：我们几位初创者的实践可以追溯到1994、1995年，当时我们都在为程泰宁院士工作。当时所形成的工作方式、习惯和标准成为我们后来共同工作的基础。创业初期，我们达成了这样的共识，既希望做一家市场化运作的公司，但又同时保持对设计品质的绝对坚持。

问：什么时候开始启用了"goa大象设计"这个名字？当时有哪些考虑？

何兼：2014年，公司启用"大象设计"这一中文名称，2018年注册英文名称"Group of Architects"，取意"一群建筑师"。从这以后，公司正式以"goa大象设计"为全称名字。选择这个名字有两点原因：首先，它相对中性，也有比较宽广的外延；其次，我们认为公司的核心是"人"，公司实践的基石在于一群什么样的人以什么样的方式共同工作，"一群建筑师"的称法传达了我们对于公司组织模式的期望。

2 设计定位与合作模式

问：对在座的三位来说，你们对自己实践的路线和类型是如何定位的？

陆皓：在早期，我们与绿城有着紧密的捆绑关系。当时绿城还没有完整的设计管理队伍，我们便被授予很大的权限，在独立的建筑师工作之外承担一部分的甲方职责，全权负责项目在设计层面的价值实现。我们的工作范围从对项目地块的评估、任务书制订一直延伸至施工材料的选定、对室内及景观细节的控制。得益于绿城对于产品品质的专注度，我们从中锻炼出全面的执行力，也养成了主动理解使用者需求、高标准把控环境品质的工作方式。这些内容被一直延续至大象设计今天的设计工作中，可以说是为我们奠定了实践"路线"的基础吧。

何兼：我们的设计实践也不能被简单概括为公建或住宅中任意一种。对于建筑设计，我们大致将其划分成两个类型：一种需要直接面向市场和大众；另一种则并不需要。大象设计的实践更关乎前者，无论是住宅、办公楼、酒店或其他领域，我们会努力去理解末端使用者的真实需求，再结合当下市场的条件予以有效的回应。相较于以简单的"公共建筑"或"住宅建筑"来划分实践路径，这样的定位方式更符合我们实践的客观特征。多年以来，我们一直将"人"的需求置于建筑项目价值系统的中心；我们相信，当我们努力实现这一核心价值之后，项目的市场价值便会作为附赠品自然而然地兑现。

问：我们每个人可能都会有心目中的榜样，比如在读书的时候想要成为柯布或路易·康那样的大师。那么在全球当代建筑设计机构的横向比较中，大象设计是否有"我们要成为谁"的对标？

陆皓：对于"成为谁"的愿望或想象是一个理想化的概念。基于大象设计的组织架构，我们的公司并不代表任何个人的理想。组成公司的成员们有着自己的个性与面貌，而公司也因此具备持续生长的不同可能性。相比于国际上很多历史悠久的设计公司，25岁的大象设计仍然处于一个很年轻的状态，我们不可避免地会开始面临代际传承的问题。回看过去，我们的实践路线由我们所选择的工作方式、运行制度所定义；而未来"要成为谁"，同样有赖于这些内容的持续建构。相比于锚定一个具体的"榜样"，这一点更为重要。

何兼：从我的角度来看，有两家设计公司是我特别关注的。一个是日建设计，另一个是SOM。日建设计对于自身建筑设计的预定目标有非常清晰的愿景，而SOM则善于从结构上解决宏观问题。在大象设计早期的设计实践中，新城市主义曾对我们产生过比较重要的影响。

问：接下来我想请你们谈谈公司合伙人之间的协调方式。在你们公司内部，如果在项目进行过程中出现分歧，一般会如何处理？

何峻：我们基本上非常稳定。假设我们每个人都成立自己的公司，各自做不同甲方各自的项目，必然会对市场认知有很大差异。但是我们的公司在创立之后的很长一段时间内，绿城一直是我们的主要甲方，这对于团队的稳定性

Interview with GOA: Embracing the Value of Design in Changes

Interviewer:
Architecture China Magazine

Interviewees:
LU Hao, President and Principal of GOA
HE Jian, Executive President and Principal of GOA
HE Jun, Executive President and Principal of GOA

1 Foundation and Development Milestones of GOA

Q: From an industry perspective, what drove the establishment of GOA, and what were the initial goals?

HE Jian: The roots of GOA trace back to 1994–95 when our founding members collaborated under Academician Cheng Taining. The work methodologies, habits, and standards we developed during that period served as the foundation for our future endeavors. In our early days, we envisioned a company that would align with market dynamics while maintaining solid design quality.

Q: When was the name "GOA" coined, and what considerations contributed to this choice?

HE Jian: The name "大象设计" was introduced in 2014 as our Chinese identity, and in 2018, we registered the English name "GOA (Group of Architects)". This marked the company's full transition to being recognized as "goa大象设计". There were two key factors that guided this naming choice. First, the name connotes both its neutral and expansive character; and second, we believe that "people" constitute the nucleus of the firm. The central value of our practice really lies in the passionate people we bring together and their collaborative synergy. "Group of Architects" accurately describes our vision for the company's organizational structure.

2 Design Positioning and Collaboration Model

Q: How would you describe the trajectory and typology that define your practice?

LU Hao: Initially, we forged a close collaboration with Greentown China, a time when they were assembling their design and management teams. During this period, we took on responsibilities that extended beyond our roles as architects. We assumed tasks typically reserved for the clients, focusing on maximizing the project's value through design.

This experience resulted in a variety of unanticipated values. For one, it honed our execution skills and cultivated a proactive approach to understanding user demands. Further, it ensured environmental excellence values instilled by Greentown China's emphasis on product quality. These practices have continuously influenced GOA's current designs and, you might say helped determine our "trajectory".

HE Jian: Categorizing our design practice as strictly "public" or "residential" would be limiting. We tend to classify architectural design into two broader categories: those that directly serve the market and the public, and those that do not. Traditionally, GOA has been focused primarily on the former only. Whether it's residential complexes, office structures, hotels, or other sectors, we prioritize comprehending genuine user needs and responding effectively to current market conditions. This positioning, rather than a rigid "public" versus "residential" division, aligns more closely with our core values. The industry has always been driven by its ability to negotiate a variety of competing values. For us, we have worked hard to maintain our focus on genuine human needs, keeping these as the cornerstone of our ethos. Perhaps we will be proven wrong one day but so far we have every reason to believe that this passionate drive to increase human value in our projects and choices will translate naturally into greater market value.

Q: We all have a role models, such as aspiring to be a master like Le Corbusier or Louis Kahn. Is there a benchmark or aspirational figure for GOA in today's global architectural landscape?

LU Hao: The idea of "becoming someone" is quite idealistic. Given its distinct organizational structure, GOA does not emulate the ideals of any individual. Our members contribute their unique personalities and perspectives, nurturing diverse possibilities for GOA's development.

At just 25 years, GOA is relatively young when compared with well-established international design firms. This youthfulness comes with the inevitable challenge of maintaining intergenerational continuity. Our journey has been forged by the methodologies and operational

有很大帮助。公司发展至今，我们多了一些多样化的见解，也是因为对于客户画像产生了不同的理解。我认为想法的多样化是好事情，只要处理好，会对公司发展有正面作用。

陆皓：这种稳定性的背后原因是多重的。其一，公司的壮大一直是通过渐进式的、小规模的人才吸纳；其二，公司的合伙人制度通过一系列的具体规则为团队合作提供了支撑。总体而言，我们今天的稳定状态得益于自公司创立以来行业市场格局与团队成长步调的契合性。当然，当时势变化，这种契合性是否能够延续还并不可知，因此我们非常关注公司运行制度、人才制度的不断建设。我认为一家优秀的公司应具备的特质之一，是公司的生命比创始人的生命更长。

何兼：一家公司是否能够凝聚的深层原因在于其所推崇的解决问题的方式是否具有长久的生命力。目前，公司正面临代际更迭，我们尤其关注公司人才梯队的培养，希望有新一代的生力军继承、延伸、迭代既有的方法论。

3 整体环境思维与建筑的综合体验

问：在你们的作品中，你们如何理解人居空间与周围环境的关系？比如就木守西溪酒店来说，居住的空间并不仅仅是物理结构，而是一系列综合体验的总和，周围的环境要素对建筑的整体品质有很大影响。

陆皓：自初创时期便开始的住宅实践令我们养成了关注环境整体的思维方式。住区总图关系的处理是一个区域级别的问题，需要综合解决一系列的序列关系，一直到确定每一栋住宅单元的位置、每一条步行道的形态。这种思维方式被自然而然地延续至我们后来所作的酒店设计中——度假空间更需要富有故事性的空间序列来吸引人，但凡有割裂之处必然影响整体性的叙事。时至今日，我们面对任何项目都会自然而然地选择这样的入题方式，以使用者在环境中的综合体验为第一考虑点。这个综合体验里面不仅包括自然的体验、文化的体验，还有人与人交往的体验。

何峻：回答这个问题需要我们回到一个更根本的问题，回到"人"本身。我们这群建筑师都有各自独特的个性，但我们都非常热爱生活，且乐于将我们对生活的体验通过建筑传达出来。大象设计过去取得成功的项目大多源于精确地捕捉、把握并传达了建筑设计与生活情感之间的联系。

何兼：我们项目的竣工照片往往优于效果图，现场的体验又往往优于竣工照片。这是因为我们的设计对象是"体验"，"体验"无法通过片段式的媒介来呈现，而在于真实现场给人的整体共鸣。在设计工作中，大家往往会划定边界，建筑、景观、室内各负其责，但在需要彼此交界的地方，常常会有缺漏。这使得整个项目的质量完全依赖于工程现场的管理水平。我们的整体环境思维在于积极填补专业之间的过渡地带。只要业主在这方面和我们达成默契，就会获得一个相对高品质的结果。

何峻：我们的触点并不局限于室内或景观这些设计的细分层面，有时也会参与建筑的策划，以争取相对理想的前置条件。以前宋卫平先生带领我们勘地时，会让我们从建筑师的角度帮他判断这块土地是否值得拿。在这种参与方式的影响下，我们偶尔也会对甲方的任务书提出挑战。

4 市场环境的应对与挑战

问：杭州对你们而言意味着什么？比如传统的江南建筑、杭州的生活方式、城市节奏或文化面貌，对你们的创作或实践有哪些隐性或显性的影响？

何兼：在文化层面，杭州有着浓郁的地方特色，可以说杭州影响了我们对于建筑在地性及文化认同命题的认知。同时，杭州对于我们的特殊性还体现在设计市场的差异上。在我们刚进入这个行业的时候，相比上海等城市，杭州的行业话语体系结构是相对自由、开放的。因此，杭州给了我们较大的空间，不仅是市场的空间，还有思维方式上的空间。

问：目前中国建筑市场已经面临着增速放缓的挑战。未来可能会有更多年轻

systems we've chosen, and their evolution will determine our future identity. For me, this is far more important than fixating on one single "role model".

HE Jian: Speaking from my perspective, I hold Nikken Sekkei and SOM (Skidmore, Owings & Merrill) in high regard. Nikken holds a resolute design vision, while SOM excels in resolving macro-level structural intricacies. In GOA's early practices, the principles of New Urbanism significantly influenced our approach.

Q: Could you share some thoughts about the collaboration dynamic among your company partners? How do you address disagreements that may arise during a project development?

HE Jun: Our team's stability has been a defining factor. Our perceptions of the market might have diverged if each of us ventured into independent firms, engaging with different projects and clients. However, Greentown China remained our central client during our formative years, imparting a steady foundation to our team dynamics. As the company expanded, different perspectives emerged due to different interpretations of client profiles. I believe whole-heartedly this diversity can be beneficial for constructive growth when effectively managed. Overall we understand our diversity as an asset rather than a challenge since we have always sought the type of conflicts which are necessary for creativity to emerge.

LU Hao: Our stability is underpinned by several elements. For one, gradual and controlled talent acquisition has helped. Additionally, our partnership system provides a robust framework for collaboration, delineated by specific guidelines. Overall, our current stability results from the long-term alignment between the evolving market dynamics and our brisk pace of growth. Maintaining this alignment hasn't always been clear and certain, as it is subject to paradigm shifts and the pulling forces of competing values. Consequently, we pay attention to the continual refinement of our operational and talent systems. In my view, the longevity of an eminent company should outlive its founders.

HE Jian: The underlying key to a company's unity lies in the enduring vitality of its problem-solving methodologies. Given the ongoing generational transitions, our current focus is on developing a talent pipeline with the hope that the next generation will inherit, expand on, and refine its mission and activities.

3 Integrated Environmental Thought and Comprehensive Architectural Experience

Q: How do you perceive the relationship between living spaces and their contextual surroundings within your works? For instance, in the Muh Shoou Xixi, the living space goes beyond physical structures; it encompasses a series of comprehensive experiences where environmental elements are significant in defining architectural quality.

LU Hao: We began cultivating an emphasis on a holistic environment from the very beginning of our residential endeavors. Managing the relationships within a residential master plan entails a regional perspective that requires a comprehensive solution to sequential relationships. This ranges from determining the positioning of each residential unit to shaping pedestrian pathways. This approach organically extended to our resort designs. Vacation spaces particularly demand narrative-rich spatial sequences to captivate visitors. Any disruption in this narrative coherence can fragment the overall experience. Presently, when approaching a project, our instinct is to prioritize users' overall experience with the environment. This all-encompassing experience includes interactions with nature and culture along with human connections.

HE Jun: To answer this question, we must return to a fundamental aspect of our work, the "people" themselves. While each architect is distinct, a common thread is the appreciation for life and gratification in expressing life through architecture. The success of our previous projects is attributed to our ability to capture and convey the symbiotic relationship between architectural design and the range of emotions inherent in life.

设计公司采取不同策略，如缩小公司规模、数字化转型、参数化设计或强调生态等，你们是否会主动调整以迎接这些挑战？

陆皓：相比于从形式逻辑入手的设计方法，我们所采用的从环境的完整性和序列性出发的方法论相对含蓄，业主对此的认知程度也有很大的差异。如何弥合这类认知差异从而建立更多共识是我们工作的挑战。

何兼：目前我们的工作方式能够适应市场的前提条件在于，我们所面对的大多是政府主导或政府交给开发商的大规模项目，我们因此能够很好地发挥处理各种整体性与协调性问题的长项。总体而言，我认为我们应当保持这样的定位，基于中国的发展进程介入一些规模化的、复杂度高的问题。诚然，如果中国市场发生巨大的变化，我们就必须面对不同的情况做出调整与适应。我们总是在与时代环境的互相作用之中修正自身实践的走向。

HE Jian: Our completed projects often transcend renderings in photographs, and the on-site encounters exceed the impact of the photos. This is because our design objective centers around the "experience", a facet challenging to convey through fragmented media but profoundly felt within tangible settings. People often compartmentalize architecture, landscape, and interiors during the design process, each holding its own responsibilities. Yet, gaps can emerge at their intersections, potentially compromising the project's overall quality. To bridge these gaps, we strive to adopt an integrated environmental approach that actively fuses these specialized domains. Clients who share this viewpoint tend to achieve notably higher-quality outcomes.

HE Jun: Our engagement extends beyond interior or landscape details; sometimes, we engage in architectural programming to establish optimal preliminary conditions. In the past, guided by Mr. Song Weiping during site visits, we assessed whether a land parcel was architecturally worth acquiring. Under such immersive influences, we occasionally feel a strong need to challenge the client's initial directives.

4 Market Environment Responses and Challenges

Q: What does Hangzhou mean to you? How do elements such as traditional Jiangnan architecture, the city's lifestyle, urban rhythm, or cultural identity influence your design philosophy, whether subtly or overtly?

HE Jian: Hangzhou bears profound cultural characteristics, teaching us the importance of grounding architecture within its local fabric while also fostering a distinct cultural identity. Hangzhou's essence also manifests through its distinct design market dynamics. During our initial steps in the industry, Hangzhou's professional discourse exhibited a refreshing openness compared to cities like Shanghai. This broader platform not only expanded our market horizons but also invigorated our creative approach.

Q: Currently, China's architectural market is undergoing a deceleration in growth. Looking ahead, emerging design firms might adopt strategies such as downsizing, digital transformation, parametric design, prioritizing ecological concerns, etc. Are you considering making proactive adjustments to tackle these challenges?

LU Hao: In contrast to design methods driven purely by formal logic, our focus on environmental integrity and sequential narrative is more nuanced, resulting in diverse degrees of client recognition. The task of bridging these perception gaps to establish a greater consensus remains a persistent challenge in our work.

HE Jian: Our current market-adaptable approach results from our engagement in substantial government-led or mandated projects. This allows us to effectively address complex coordination issues. Overall, I support maintaining this stance, immersing ourselves in expansive, multifaceted issues that align with China's developmental trajectory. While it is true that we must be ready to adapt to significant changes within the domestic market, our fundamental philosophy remains consistent: iteratively refining our practice in response to the ever-changing landscape.

goa 大象设计总部
GOA Headquarters

项目地点：浙江省杭州市
建筑面积：16,900m²
设计 / 竣工：2018/2020
Location: Hangzhou, Zhejiang
Floor Area: 16,900m²
Design/Completion: 2018/2020

办公大堂 / Lobby

Headquartered in OōEli, Hangzhou, the interior design of GOA is done by its own team. The interior design integrates the architect's daily work scenes inside, while following the overall style and the rigorous modulus system of architectures in the park, expressing a delicate sentiment and humanistic quality.

At OōEli, fair-faced concrete construction is a prominent design element. The exposed one-piece concrete texture with a silky finish on the ground floor lobby serves as both the structural members and a centerpiece of the interior space.

The heart of the GOA headquarters comprises two full-height atriums spanning the 3rd through 7th floors to form a stepped library that connects all the open workspaces. The lower atrium from the 3rd to 4th floor features platforms that create multiple small gathering areas to foster creative collaborations. The upper atrium from the 5th to 7th level is outfitted with projection screens, sound facilities, and specific lighting systems that allow the space to be converted into an auditorium for over 100 people. The 8th floor conference center offers ten meeting rooms of varying sizes for groups of up to 30 people and three large conference rooms that can combine to host up to 150 people. On the southern side of the conference center, a shared leisure lounge with an outdoor terrace offers the best landscape viewing perspectives. The design weaves the essence of OōEli's "Natural and Green Core" concept into the interior design, resulting in a comfortable and luminous atmosphere with enhanced spatial transparency.

The "g MateriaLab" is over 200m² and contains over 1,000 material samples. The GOA-developed data platform compiles information on each material's specification, origin, manufacturer, and images of related projects. The designers can quickly access all material-related data by scanning codes with their mobile devices.

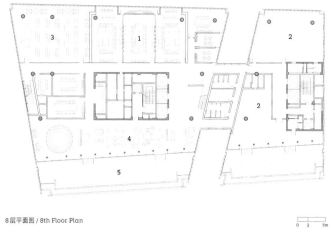

8层平面图 / 8th Floor Plan

1	会议室	1	Meeting Room
2	办公区	2	Office
3	g料间	3	g MateriaLab
4	g空间	4	g Space
5	露台	5	Terrace

goa大象设计总部位于杭州天目里综合艺术园区内，室内设计由大象设计团队独立完成。设计遵循了园区整体的建筑风格与模数系统，并于此基础之上融入对建筑师日常工作场景的设想，使得整体空间呈现细腻的情感和独特的人文气质。

清水混凝土结构的露明处理是园区建筑最重要的特色之一。大象设计总部一楼大堂延续了园区建筑的整体风格，细腻的清水混凝土墙面一次浇筑成型，结构构件亦是室内空间的主角。

从3至7层，两座通高中庭将所有开放式办公区域联系起来，形成了一座垂直的阶梯图书馆。3至4层中庭利用高差形成多个小型讨论区，空间共享且互不干扰；5至7层中庭配备了投影、幕布、音响设施、灯光，可以随时化身为阶梯教室，举行百余人规模的讲座活动。

位于8层的会议中心设有10间不同规模的会议室，能够满足1—30人的会议需求。其中三间大会议室还可以连通，形成可最多容纳150人同时使用的开放空间。会议中心南侧是公共景观露台和开敞的休闲区，这里拥有楼内最佳的观景视野。设计师希望将整座园区的重要概念——"自然的绿色内核"延续到室内设计之中，使更多景观、阳光、新鲜空气渗透进室内，强化空间的透明性，给人以舒适明朗的感受。

材料实验室被命名为"g料间 / g MateriaLab"，面积超过200m²，收藏了超过1000种建筑、室内的材料。大象设计自行开发的数据平台收集了每种材料的规格、产地、厂家以及应用该材料的项目实景照片。建筑师可以直接通过手机扫码查询，便捷高效。

5至7层中庭：g书馆 / 5-7F Atrium: g Library

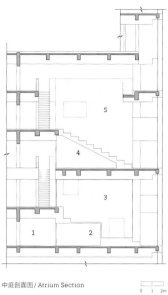

中庭剖面图 / Atrium Section

1　茶水间　　1　Pantry
2　储藏室　　2　Storage
3　3至4层中庭　3　3-4F Atrium
4　设备间　　4　Utility
5　5至7层中庭　5　5-7F Atrium

3至4层中庭 / 3-4F Atrium

027

g 空间 / g Space

g 料间 / g MateriaLab

会议等候区 / Waiting Area

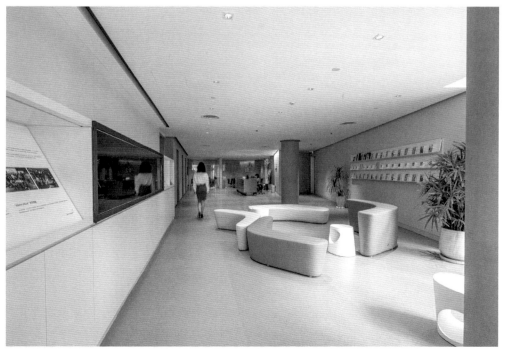

接待区 / Reception

For

Mixed-use

混合使用开发

上海西岸金融城 G 地块
Shanghai West Bund Financial City - Plot G

鹿山时代
Lushan Times

深圳万创云汇 01—03 地块
Shenzhen Vanke Cloud Gradus - Plot 01-03

上海长风中心
Shanghai Changfeng Center

重庆启元
Century Land Chongqing

恒力环企中心
Hengli Global Enterprise Center

杭州西动所上盖及周边区域综合开发
Hangzhou West High-speed Train Maintenance Base Superstructure

望江中心 TOD
Wangjiang Center TOD

滨耀城
Colorful City

天目里
OōEli

上海西岸金融城 G 地块
Shanghai West Bund Financial City - Plot G

项目地点：上海市徐汇区
建筑面积：97,500m²
设计／竣工：2020／—
Location: Xuhui, Shanghai
Floor Area: 97,500m²
Design/Completion: 2020/-

The West Bund Financial Hub, once a key Shanghai transportation, logistics, and manufacturing hub, is located in Xuhui District. As the first construction zone, the G plot combines residential, cultural, and artistic attributes into a living entity that propels the development of future lifestyles. The Shanghai South Railway Station Eight-line Warehouse and Platform, which once served as a pivot for water-to-land transshipment, is a historic preservation on the site. Its history dates back to the early 20th century.

The design revolves around the revitalization of the historic station. Under a wooden-brick structural system and a three-span-gable roof, the entire station structure extends 180m east-west across the site and functions as a defining spatial feature. The design transforms the existing station into a cultural landmark. The station's west wing remains in place, while the middle and east wings are pushed out to form a street interface that preserves the linear spatial form. Two historical axes connecting various spatial nodes are established by redesigning the station. New structures, including shops, restaurants, and exhibition spaces, integrate with the renovated station, enclosing an attractive garden.

A fusion of traditional and modern elements gives the architecture a captivating charm. The roofs morph from gable to free-form canopies, defining the art garden's entrance, the art bookstore's front elevation, and the apartment lobby. This large-span three-dimensional structure creates a harmonious symphony. The original masonry from the relocated wings intertwines with glass bricks to form the new facade, resulting in an interplay of materials that weaves a delicate tapestry of time.

Overlooking the garden from a gentle stance, the residences and rental apartments with sleek balcony edges seems to conduct a dialogue between historical echoes and modern aesthetics.

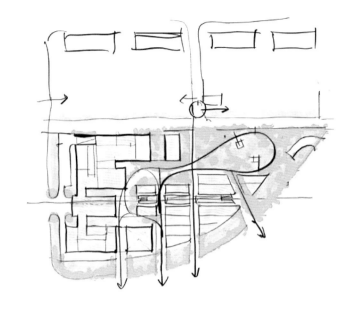

上海西岸金融城所在的徐汇滨江地区是上海近代最主要的交通运输、物流仓储和生产加工基地，也是中国近代民族工业的摇篮之一。首发建设的G地块，计划建成集居住、文化、艺术于一体的混合式街区，是该区域未来生活的发生器。场地内既存的一般历史建筑上海南站八线仓库站房及站台建于1957年，曾是水陆联运的重要枢纽，其历史可追溯至20世纪初。

设计发端于对老车站的再生策略。东西向贯穿场地的站房和站台建筑均为砖木混合结构，三跨人字坡顶所覆盖的180m超尺度线性长廊是其显著的空间特色。方案以去粗存精的态度将其改造为区域中的文化标志物：西段长廊被原位保留，中段与东段则在整体移位后成为对外展示面，延续三跨人字坡顶的特征。以此形成的两段中轴线成为两条历史的轴线，串联起各空间节点。集合零售商业、餐饮、艺术展示等功能的新建建筑与改造后的车站共同围合出宜人的花园，迎接城市四面八方的访客来此漫游、休闲。

建筑形态的魅力来源于新与旧的并置与融合。局部屋面由传统的人字坡渐变为极具现代感的波浪形屋面，勾勒出艺术花园的入口、艺术书店的正立面及公寓的门厅，大跨度的三维起翘结构协奏出美妙的乐章。移位后的车站局部立面采用灰砖与玻璃砖拼接的工艺，利用质感的渐变实现新与旧共生的肌理。

花园之上，带有流线形阳台的公寓及住宅楼宇以优雅、轻盈的姿态俯瞰着江水，仿佛在与历史时空里的城市记忆对话。

034　　混合使用开发　　For Mixed-use

鹿山时代
Lushan Times

项目地点：浙江省杭州市
建筑面积：173,000m²
设计／竣工：2013/2020
Location: Hangzhou, Zhejiang
Floor Area: 173,000m²
Design/Completion: 2013/2020

As a significant milestone in district development located at the entrance of Lushan New City, Fuyang of Hangzhou, the Lushan Times is a city-level complex that integrates commercial, office, and hotel spaces.

The design features a unique dual-ring spatial structure, with a sunken spiral square nestled within a curvaceous triangular contour, forming courtyards with distinct themes stretching aligned to the two boundaries. Multiple volumes rise from the perimeter podiums. This enclosing gesture enhances and expands the riverfront vista while preserving a pure and cohesive architectural expression.

The "New Urban Parlor" concept aims to promote spatial convergence by transforming limited architectural volumes and public spaces into a hub for the city's vitality. The terraced outdoor platforms of the lower commercial space elevate the ground value and offer multiple landscape perspectives. A 60-meter-wide spanning structure forms the main entrance, creating a concave plaza facing the river to invite commercial circulation and introduce a river view for the apartments at the back.

The building's curved facade is enlivened with rhythmic thickness variations on aluminum alloy profiles. Viewed from different distances, these horizontal components produce a dynamic interplay of light and dark that resembles the mountain silhouettes, just as the undulating natural skyline across the river casts a shadow.

Since its operations, the Lushan Times has become a catalyst for urban vitality by fostering functional synergy, celebrating public activities, and embracing scenic value. It breathes a new life into the Fuyang Lushan New City, creating vibrant urban settings.

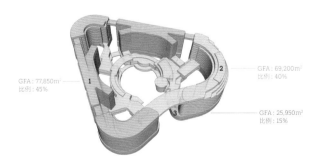

1 办公	1 Office
2 酒店	2 Hotel
3 商业	3 Commercial

本项目位于杭州富阳鹿山新城的入口地带，是一座集商业、办公和酒店于一体的城市级综合体，也是新城建设进程中的重要里程碑。

设计采用独特的双环空间结构，在弯曲的三角形外轮廓中嵌套下沉的圆形广场，双环之间勾勒出若干氛围各异的边庭。外围合的策略在保证形象纯粹性的同时实现了看江视野的最大化与多元化。

建筑师提出"新城会客厅"的构想，着力加强空间的聚合效应，使有限的建筑体量与公共活动内容结合，发挥出更为显著的、城市层级的贡献。下部商业通过阶梯式的立体室外平台设计，将地面价值逐渐引入上方，同时也创造出多标高的看江视角。在面向江面一侧，建筑退让红线，形成内凹的城市广场，60m长的挑空洞口对商业人流形成引导，并为后排公寓引入江景。

立面采用宽窄不一的横向线条构建曲面的韵律，从不同的距离观看，曲面上具有弧度的横向铝合金构件在光线的照射下反映出亮暗的区隔，如同具有了山峦轮廓的投影。

自运营以来，鹿山时代通过促进功能协同、激发公共活动和拥抱景观价值，成为城市活力的催化剂，给鹿山新城的市民带来了崭新的生活载体。

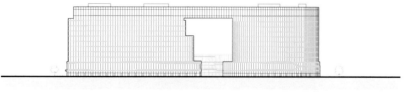

沿街立面图 / Street-front Elevation

深圳万创云汇 01—03 地块
Shenzhen Vanke Cloud Gradus - Plot 01-03

项目地点：广东省深圳市
建筑面积：224,900m²
设计 / 竣工：2021/—
Location: Shenzhen, Guangdong
Floor Area: 224,900m²
Design/Completion: 2021/-

The project is situated in the northeastern corner of the central Futian District, Shenzhen, surrounded by the scenic beauty of Shenzhen Central Park, Lianhuashan Park, and Bijiashan Park. Its location in the Meicai area predominantly serves as a science and technology innovation hub. As the first redevelopment plot in the area, the project prioritizes the strategic integration of diverse functions while emphasizing public accessibility to deliver a comprehensive settlement that integrates work, residence, and recreation within the high-density urban fabric.

A 3D-integrated circulation system reinforces the connection between the architecture, adjacent mountains, and the urban park to the south. The ground level incorporates urban infrastructure and logistics. The second level contains a pedestrian system that attaches to the external civic corridor network and provides a further public interface between the architecture and the city. The podium introduces verdant vegetation to roof terraces, forming an urban green valley.

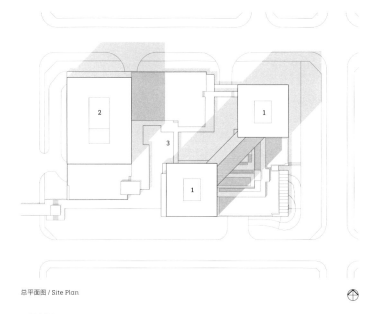

总平面图 / Site Plan

1　配套宿舍　　1　Dormitory
2　研发办公　　2　R&D Office
3　城市公共通道　3　Urban Public Corridor

The project seeks to create a dynamic, inclusive community accommodating diverse living arrangements. The podium, which is shorter than 24m tall, provides various public services and amenities, such as R&D offices, cultural event venues, and senior nursing facilities. On the podium roof, a 2,500m² outdoor communal activity area, accessible 24/7, directly connects to the pedestrian system and indoor sports club.

Three rising towers of varying heights trace an undulating urban skyline. The initial scheme incorporates a stunning, transparent infinity pool spanning two adjacent residential towers, featuring a spectacular focal point of the complex.

本项目位于深圳福田区的中心区东北角，周边环绕着中心公园、莲花山公园、笔架山公园等城市公园。其所在的梅彩片区以创新科技、信息技术及专业服务为主要定位。作为片区更新的首发地块，设计旨在建立集工作、居住和生活为一体的综合型聚落，在高密度开发的前提之下达成紧凑混合、开放共享的目标。

设计以多层次立体流线加强建筑与临近山地和南侧城市公园之间的联系。建筑底层纳入了城市基础设施——公交车首末站。二层南北、东西两个方向的公共步行通道与外部市政连廊体系相接，是建筑接驳城市的第二个公共基面。裙房上部的层层退台引入绿色，形成城市绿谷般的公共交往空间。

本项目旨在创造一个充满活力的包容性社区，以适应多样化的生活需求。24m高度以下的裙房设置创新型研发办公、文化活动、社区服务和老年托养等多种公共服务设施。2500m²的社区体育活动场地被置于研发办公裙房的屋顶，24h对公众开放，可由首层地面及二层城市公共通道直达，与室内的运动健身场地联动。

裙房上方，三座高度不同的塔楼构成错落有致的城市天际线。初期方案还曾构想于两栋邻近塔楼之间建造悬空的玻璃泳池，塑造整组建筑群落的标识性亮点。

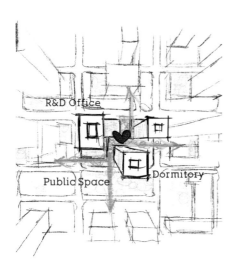

上海长风中心
Shanghai Changfeng Center

项目地点：上海市普陀区
建筑面积：359,800m²
设计/竣工：2011/2021
Location: Putuo, Shanghai
Floor Area: 359,800m²
Design/Completion: 2011/2021

The Shanghai Changfeng Center is situated between the middle and inner ring of Putuo District, adjacent to Changfeng Park. It was born out of a vision to enhance this well-established region. This comprehensive mixed-use development consists of three phases, including programs such as commercial, Grade-A office, headquarters office, and apartments.

By urban regulatory plans, the irregular site is traversed by a green belt incorporated into the design as a green core and primary pedestrian axis connecting the three clusters. From the public green belt to individual courtyards and residential green spaces to commercial plazas, a cohesive public space system promotes the integrity of the entire block.

The central green belt, approximately 30m wide, is expanded into a 110m x 90m garden to the north. A smooth transition is introduced between the constructed elements and the natural beauty by dividing the podium surrounding the garden into smaller volumes using cutting, pixelating, and setback techniques. The headquarters office adopts a grid pattern comprising 10 four-story point structures with rooftop terraces. This matrix surrounds a serene rectangular garden, seamlessly merging the architectural cluster and landscape.

The commercial cluster covers an area of approximately 40,000m², catering primarily to the block's office workers and nearby residents. In contrast to enclosed shopping malls, the project features an open street layout that connects seamlessly to the surrounding community and offers a variety of leisure options for office workers. The circulation revolves around the "activity plaza - meandering pedestrian street" levels. The entire commercial block is open 24 hours a day, functioning as an extension of the urban streets.

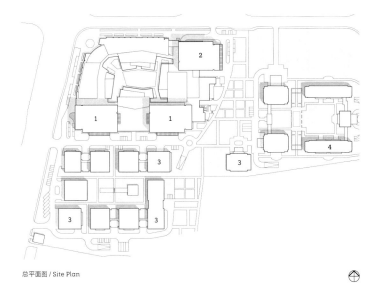

总平面图 / Site Plan

1	商业	1	Commercial
2	甲级办公	2	Grade-A Office
3	总部办公	3	Head Office
4	公寓	4	Apartment

本项目位于上海普陀区，地处中环与内环之间，毗邻长风公园，它的诞生源于对这个成熟区域进行品质整体提升的愿景。这一综合性开发项目分三期完成，包括商业、甲级办公、总部办公和公寓。

根据规划，不规则的场地被一条城市绿带穿过。城市绿带成为了整个街区的绿色交通枢纽，联系着三个组团，也是整个街区的绿核。从城市绿带到内向庭院，从宅间绿地到广场，公共空间被作为一个完整体系加以考虑。

宽约30m的绿带于中段经空间放大扩展为110m×90m的方形花园。方形花园的四周，设计对裙房建筑体量进行"化整为零"的处理，以体量的切分、像素化、退台等设计手法，让建筑物与绿带之间的边界变得更为柔软。总部办公组团由10座带有屋顶露台的四层点状小楼组成，以十二宫格秩序围合出一个矩形的静谧花园，整组建筑与地面景观融合在一起。

商业组团的体量约40,000m²，主要服务于街区内部办公人群和周边居民。相比于封闭式购物中心，项目所采用的开放式商业街区布局更亲近周边社区，也满足了上班族一天内各个时段的购物休闲需求。动线围绕"活动广场一迴游式步行街"的层次展开，整个商业街区24h全时开放，是城市街道的延伸。

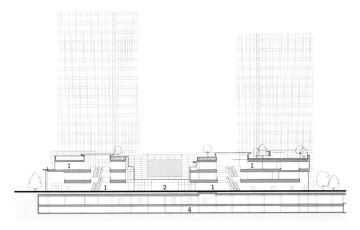

剖面图 / Section

1	商业	1	Commercial
2	庭院	2	Courtyard
3	屋顶花园	3	Roof Garden
4	机动车库	4	Parking

重庆启元
Century Land Chongqing

项目地点：重庆市江北区
建筑面积：175,400m²
设计 / 竣工：2020/—
Location: Jiangbei, Chongqing
Floor Area: 175,400m²
Design/Completion: 2020/-

The Century Land sits in the heart of the largest commercial district in southwest China, Guanyinqiao, Chongqing, adjacent to Jialing Park. It is a comprehensive urban development featuring a mix of commercial, office, and high-rise residential buildings.

The design fuses the taste of city life within a particular regional context. It divides the area into multiple platforms of differing heights to accommodate the nearly 13-meter elevation change from the site's southern to the northern border. Gestures of stacking, dropping, and staggering generate clusters of dispersed volumes within an organic architectural form, imbuing the distinctive "Mountain City" morphology with a contemporary reimagining. The cascading terraces transform the site's topographical challenges into a collection of rooftop gardens, creating an enticing slow pathway that ascends and winds through the landscape. Meanwhile, the underground commercial street directly connects to the metro station, assuring rapid and efficient circulation for heavy pedestrian flow traversing the various land parcels. Consequently, this 3D-integrated pedestrian system integrates functional and landscape values, maximizing TOD benefits while preserving the vitality of the urban core.

To navigate the intricate road networks, commercial and residential volumes feature filleted corners to avoid sharp edges that point toward the streets. Such an approach is applied to the residential facade to accentuate visual integration. Celebrating the contemporary aesthetic of a flourishing city, the dynamic horizontal rhythm of the facade is complemented by continuous vertical lines that enhance the upward movement of the residential volumes. In addition, the commercial facade is embellished with glass elevators and perforated panels to enrich visual expressions.

The project accomplished a collaborative and cohesive approach from the early stages, incorporating interior, landscape, curtain wall, and lighting. This comprehensive strategy results in integrated design practice for mixed-use development projects.

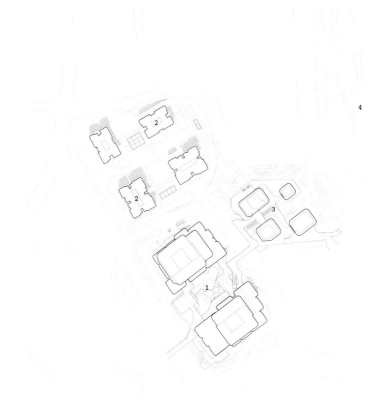

总平面图 / Site Plan

1	商办综合体	1	Commercial Complex
2	住宅	2	Residential
3	商业街	3	Commercial Street
4	嘉陵公园	4	Jialing Park

本项目位于中国西南地区最大商圈重庆观音桥商圈的核心区，毗邻嘉陵公园，是一个包括商办综合体、商业街和高层住宅的混合使用开发项目。

设计试图在这个核心商圈刻画山城的独特魅力——"立体空间"之中的"市井滋味"。方案将基地划分成几个不同标高的台地，以适应场地从南到北的近13m的高差变化；通过堆叠、跌落、错动的手法，完成对山城传统街区的现代演绎。层叠的"梯田"将场地的地形挑战转化为屋顶花园的集合，拾级而上的商业动线盘旋于花园景观之中，为人们带来饶有趣味的慢速游逛体验。同时，与地铁直接相连的地下商业街则承担"快速"通行流线的作用，在满足各个地块之间的大规模人流通达需求的同时，保留蓬勃的生活气息，充分发挥核心商圈的TOD属性。

由于周边道路的朝向关系复杂，商业和住宅体量都采用圆角设计，以避免产生面向道路的尖锐边缘，这一设计语言也被应用于住宅的立面。住宅立面的水平线条律动中，连贯的竖向线条切分出挺拔高耸的建筑体量，建筑的飘逸形象耦合着山城的时尚风貌。商业立面局部点缀景观电梯、特色穿孔板，在视觉上提供多元化解读的可能性。

项目自设计前期即秉承各专业紧密协同的理念，对建筑、室内、景观、幕墙、照明等专业的设计想法予以高度融汇，是混合使用开发类项目的一次一体化设计实践。

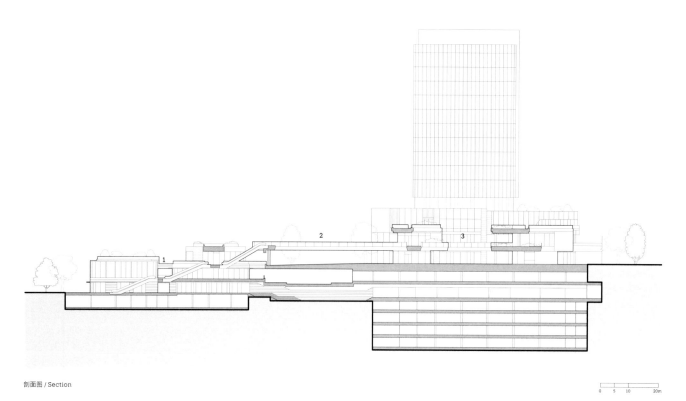

剖面图 / Section

1 商业街区　1 Commercial Street
2 空中连廊　2 Sky Bridge
3 商业中庭　3 Commercial Atlrum

恒力环企中心
Hengli Global Enterprise Center

项目地点：江苏省苏州市
建筑面积：1,277,000m²
设计 / 竣工：2020/—

Location: Suzhou, Jiangsu
Floor Area: 1,277,000m²
Design/Completion: 2020/-

052 混合使用开发 For Mixed-use

The project is located in the eastern core of Suzhou Bay Avenue, 1.8km away from Taihu Lake. It seeks to harmonize with the spatial texture of Taihu New Town and deliver a healthy, vibrant, efficient urban block that combines commercial, office, research and residential facilities.

The site is divided into three functional clusters: north, central, and south, each embracing a central garden. These clusters are interconnected through aligned axes to form a spatially cohesive relationship. Additionally, the design optimizes the scale of passages between buildings to facilitate the entry of natural light and ventilation while guaranteeing appropriate spatial depths. Different functions synergize vertically and horizontally within the project. The commercial podium offers a vibrant experience and provides essential support for the upper office and residential spaces, ensuring the sustained vitality of the entire area.

The site's western edge features two 260m-tall symmetrical towers. These iconic twin towers emphasize the urban silhouette and mark the starting point of a green axis that leads toward Taihu Lake. Anchored by the twin towers, the surrounding buildings create a stable "山" (mountain) shaped structure, gradually decreasing in height towards both sides and eventually blending into the lakeside skyline.

Multiple shared spaces are incorporated into this 260m-tall tower to prioritize a human-centered experience in high-density development. The tower's central area features a nearly 50m-high atrium with skylights, allowing natural light to illuminate the office spaces. A sky club and an observation gallery tower's pinnacle offer panoramic views of Taihu Lake, creating a scenic environment for leisure strolls. The tower facade employs the Suzhou traditional Kesi Technique, known for its "passing warp thread and cutting weft thread" skill weaving skill, emphasizing the sleek verticality of "warp" with the underlying horizontal flow of "weft."

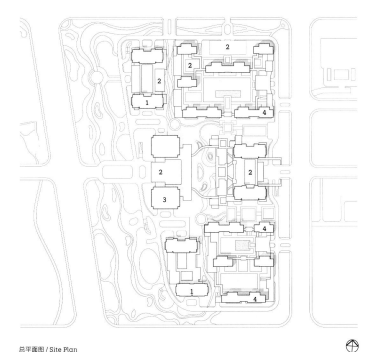

总平面图 / Site Plan

本项目位于苏州吴江太湖新城苏州湾大道东侧的核心位置，距离太湖1.8km。设计旨在遵从太湖新城整体空间秩序的同时，利用建筑的组合关系塑造宜人的公共空间，通过商业、办公、科研、公寓等功能的复合塑造健康、活力、高效的城市街区。

场地被划分为北、中、南三个组团，各组团均围绕中心花园展开，以对齐的轴线相互连接。此外，设计在保证建筑合理进深的前提下，尽可能满足建筑之间的开阔间距，以引入自然采光和通风。不同功能在竖向和水平方向发挥了协同作用。商业裙房塑造鲜活的街区体验，也为上部的办公、公寓空间提供充分的生活支持，确保整个区域的持续性活力。

场地以西，是高楼林立的繁华景象。作为项目形象的制高点，两栋对称的260m超高层塔楼被布置于场地西端，面向远处的湖景与近处的城市绿带。对称双塔的形象如同门阙凸显于城市天际线中，为通往太湖的绿轴塑造出空间的起点。以双塔为中心，项目楼宇高度向两侧呈渐低之势，以"山"字形的稳重姿态融入滨湖天际线中。

为提升高密度建筑的人性化体验，设计于260m超高层塔楼中设置多处共享空间。高区中央设有近50m挑高的超高中庭，顶部天窗为办公区域引入自然光线；塔冠处设云中会所及观景长廊，允许人们在漫步游逛间享受太湖风景。塔楼立面化用苏州缂丝传统工艺中的"通经断纬"技法，强调作为"经线"的竖向构件的流畅性，作为"纬线"的横向构件则铺垫于后。

1	商务办公	1	Office
2	商业	2	Commercial
3	科研办公	3	R&D Office
4	服务型公寓	4	Serviced Apartment

杭州西动所上盖及周边区域综合开发
Hangzhou West High-speed Train Maintenance Base Superstructure

项目地点：浙江省杭州市
建设规模：410,000m²
设计／竣工：2022／—
Location: Hangzhou, Zhejiang
Floor Area: 410,000m²
Design/Completion: 2022/-

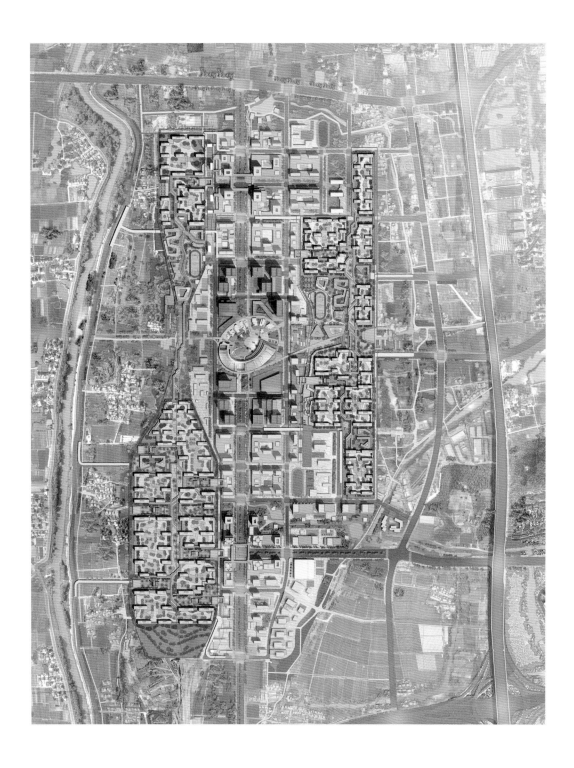

054　混合使用开发　For Mixed-use

Serving as a prototype for future communities, the project is a comprehensive transit-oriented development driven by rapid transit systems.

The Hangzhou West High-speed Train Maintenance Base is a crucial supporting facility for the Hangzhou West Railway Hub, located 4km northwest of Hangzhou West Railway Station. It is China's first high-speed train maintenance base to incorporate transit-oriented development. The master plan encompasses a mixed-use community spanning 231,700m² above the maintenance base, offering residential, commercial, cultural, and educational facilities for approximately 35,000 residents. Upon completion, this visionary "city in the sky" will be China's pioneering urban complex cluster, integrating rapid transit and dual-rail linkage with the Cangqian Depot TOD.

Typically, railway infrastructure is positioned as an isolated node away from urban centers. However, this project aims to maximize the potential of the superstructure platform and utilize land resources in a three-dimensional manner. The objective is to transform railway infrastructure into beneficial components for urban life. This winning proposal is rooted in the concept of a "boundless and symbiotic vertical city," creating an integrated and efficient three-dimensional compound community. It incorporates three metro stations and four community terminals, facilitating connectivity between various zones, including super high-rise landmarks, sunken plazas, vertically-oriented community blocks, and terraced recreational and commercial districts.

The design effectively addresses the challenges of elevation differences, vehicular diversion, pedestrian connectivity, public transit conversion, equipment layout, and urban interface. The commercial district and community terminals serve as dynamic hubs, attracting activity and creating a lively atmosphere that blends commerce and artistry within the vicinity.

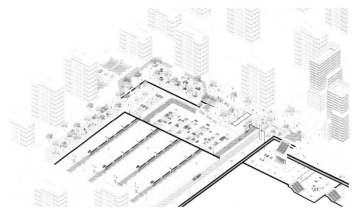

概念轴测图 / Conceptual Axonometric

本项目呈现了由轨道交通快线引领的超级TOD集群以及具有示范作用的未来社区样板。

杭州西动车运用所位于距杭州西站西北方向4km处，是全国首个进行上盖开发的高铁动车运用所，也是杭州西站枢纽的重要配套设施之一。根据规划，动车运用所上方通过约231,700m²的上盖板承载一座集居住、商业、文化、教育等功能为一体的混合社区，形成一座宛如"天空之城"的城市综合体，容纳近3.5万人居住生活。项目的愿景还包括建成后与地铁仓前车辆段上盖共同组成国内首个双铁联动、轨道交通快线引领的城市综合体集群。

传统铁路基础设施往往独立于周边环境，成为一个孤立的节点。动车运用所上盖的设计重点，在于充分利用"上盖"平台，立体化利用土地资源，将铁路基础设施转化为有益于城市生活的积极因素。中标方案以"榫卯时空，无界共生"为设计概念，依托3个地铁站点及4个邻里车站，为整片区域设想城市未来空间驿站、榫卯时空阳光广场、生活梯田漫步街区三大特色场景，通过创造盖上及盖下一体化无边界空间，实现高效利用的复合型社区。

方案集中解决上下板高差、交通分流、慢行系统连接、公交系统转换、设备管线穿越、盖板界面形象等问题，使商业街与邻里车站成为吸引人流的引擎，为周边居民提供活跃的商业与艺术氛围。

1	动车所	1 High-speed Train Maintenance Base
2	地铁站	2 Metro Station
3	办公	3 Office
4	板下商业	4 Lower-deck Commercial
5	板上商业	5 Upper-deck Commercial
6	邻里车站	6 Community Terminal
7	社区中心	7 Community Center
8	体育活动场地	8 Sports Field

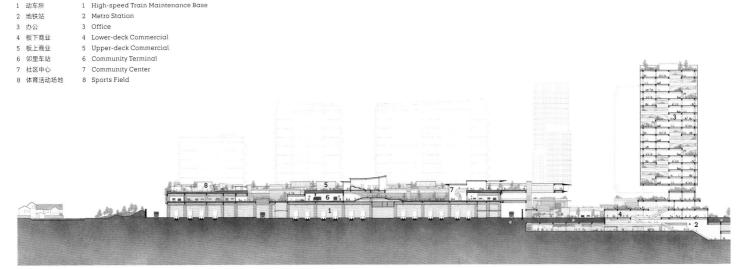

概念剖面图 / Conceptual Section

望江中心TOD
Wangjiang Center TOD

项目地点：浙江省杭州市
建筑面积：49,000m²
设计 / 竣工：2019/—
Location: Hangzhou, Zhejiang
Floor Area: 49,000m²
Design/Completion: 2019/-

056　　混合使用开发　　For Mixed-use

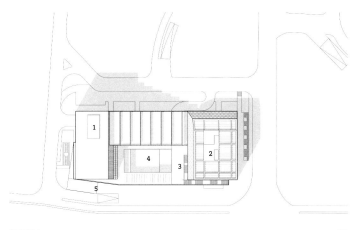

总平面图 / Site Plan

1 LOFT办公　1 LOFT Office
2 办公　　　 2 Office
3 商业　　　 3 Commercial
4 下沉庭院　 4 Sunken Courtyard
5 地铁出入口 5 Metro Entrance

Wangjiang Center TOD is situated at Wujiang Road station of Metro Line 1 in downtown Hangzhou. Due to its proximity to Qianjiang New City, it serves as the gateway to the future financial and technology axis. Under a controlled FAR of 3.8, the project achieves a symbiosis between an urban complex, a rail transit line, and compact communities under restricted conditions, resulting in a high-quality development that combines commercial, office, and residential programs.

The design utilizes a transfer truss to suspend the main tower structure above the existing metro entrance and ventilation pavilions to address the limited construction land area. The commercial podium incorporates the metro station's entrance and a large-scale farmer's market as a hub of social connection and food supply for nearby communities. A sunken courtyard at the station's exit marks the beginning of public spaces. It connects to stairs and podium terraces and guides visitors along a spiraling path to a sky lounge between the commercial and office sections, inviting them to unwind and appreciate the panoramic vistas.

The building is designed with a beveled V-shaped volume to ensure ample sunlight access for itself and neighboring residences. Its verdant terraces and roof gardens provide a fifth facade that, along with the lush sunken courtyard, improves the urban microclimate of a high-density district.

As one of the limited Chinese TOD initiatives that convert existing metro facilities, the Wangjiang Center is instructive for urban revitalization and similar developments in future high-density built environments.

望江中心TOD是杭州地铁1号线婺江路站的上盖建筑，它毗邻钱江新城，是未来婺江路金融科技轴线新的地标门户。建筑地处开发饱和、人口密集的城市核心区位，规划容积率为3.8。项目实现了城市综合体与轨道交通线路、既有城市社区之间的共生，于限制条件中完成集商业、办公和住宅于一体的高质量开发。

设计利用转换桁架将主塔结构悬挂在现有地铁入口和通风亭上方，以适应紧张的用地条件。商业裙楼中纳入的地铁站出入口和大型农贸市场，为周边居民提供了社会交流的窗口和食品供应的保障。地铁站出入口所衔接的下沉庭院是建筑公共空间序列的起点，它连接公共楼梯和裙房平台，引导游客沿着螺旋上升的路径到达办公和商业之间的空中城市客厅观景休憩。

在建筑自身及周边住宅的日照要求限制下，建筑被斜切出V字形，层层叠退的景观露台顺势成为造型的鲜明特色。塔楼的退台绿化和裙楼的露台花园共同构成建筑的第五立面，与下沉庭院景观一起，改善着高密度城市中心的微气候。

作为国内为数不多的对现存地铁设施进行改造的地铁上盖项目，望江中心对未来高密度建成环境下的城市更新和TOD开发具有一定启示意义。

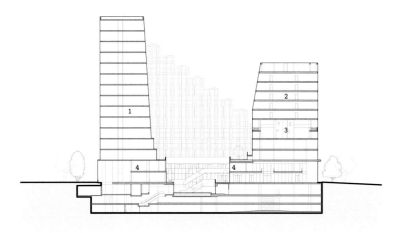

剖面图 / Section

1　LOFT办公　　1　LOFT Office
2　办公　　　　2　Office
3　城市客厅　　3　City Lounge
4　商业　　　　4　Commercial

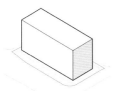

01　基础体量
　　Base Volume

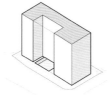

02　设置中心庭院，对城市开放
　　Inscribe to Create Sunken Courtyard

03　根据日照斜切体量
　　Cut to Maximize Solar Access

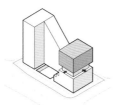

04　根据功能分配关系，调整建筑形体
　　Distrubute Programs

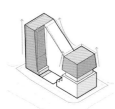

05　多角度推敲建筑比例
　　Modification

06　设置丰富的漫游动线，串联共享空间
　　Nested Circulation Connects Public Areas

滨耀城
Colorful City

项目地点：浙江省杭州市
建筑面积：262,600m²
设计 / 竣工：2019/2023
Location: Hangzhou, Zhejiang
Floor Area: 262,600m²
Design/Completion: 2019/2023

Located in the heart of Linping New City, Yuhang District, Hangzhou, Colorful City is a key element of the Donghu Riverside skyline, bordered by Donghu Park and Linping Grand Theatre to the south. The site is divided into two by Xuehai Road, with a metro line running through the middle of the east lot, creating a 30m-wide construction-free zone on the ground. Despite these limitations and challenges, the design creatively incorporates the functional requirements of the highly mixed-use program while providing a delightful spatial experience.

At the site's southwest corner, a 150m-tall tower provides sweeping views of Donghu Park. The tower's lower section is dedicated to offices, while the upper part is a hotel. On the north of the west lot are four residential buildings ranging from 80-100m in height and with a few low-rise commercial spaces, providing convenient access to services for the entire lot. The area above the metro line becomes a linear park, with a 75m-high office building at its north end. South of the east lot is designated for commercial and office buildings less than 20m in height, with a gradual decrease in height from north to south and west to east to guarantee that each building has an unobstructed view of the south-facing landscape. This configuration also creates a vibrant, sunny garden as an active communal space.

The north-south axis divides the supertall tower into two slightly asymmetrical sections, creating an appealing visual effect reminiscent of a partially unrolled scroll adorning the street. This concave and convex architectural design mitigates the tall building's imposing stature and creates a distinct sense of individuality. The tower is accentuated by the 270° oversized curved glass at the corners, creating a captivating visual journey for hotel guests and office occupants. The facades of the low-rise commercial block exhibit a contemporary Chinese aesthetic, with large horizontal windows serving as commercial displays that combine modern elements with classic influences.

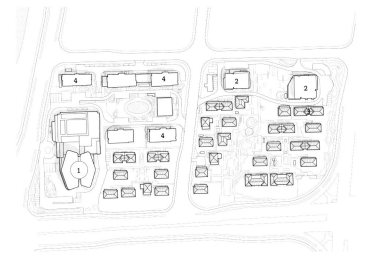

总平面图 / Site Plan

1	酒店及办公	1	Hotel & Office
2	办公	2	Office
3	办公及商业	3	Office & Commercial
4	住宅	4	Residential

滨耀城位于杭州余杭区临平新城核心地带，南侧与东湖公园及临平大剧院相接，是东湖沿岸天际线的重要组成部分。场地被雪海路分隔为东西两个地块，东地块之中地铁斜穿而过，在地面形成30m宽的非建造地带。设计的目标是将场地内的不利因素转化为有利因素，在保证各混合业态功能需求的同时，创造出舒适宜人的使用体验。

150m的超高层建筑被设于场地西南角，可以纵览整个东湖公园的景观，塔楼低区为办公，高区为酒店。西地块北侧为4座80m至100m高的住宅楼栋，底部配有少量高度20m以内的商业设施，为整个地块提供便捷的服务。地铁线路上部被规划为线性公园，其北端设有75m高的办公楼，其南侧为高度20m以内的商业及办公群落。由北向南、由西往东，建筑的高度逐渐降低，这保证了各建筑单体的南向景观视野，也使得南侧开阔区域成为一个阳光充足、景观宜人的商业花园。

超高层塔楼沿南北向中轴被切分为大体对称的两部分，如同一册微微展开的书简矗立于街道旁。凹凸有致的形态既弱化了高层建筑的压迫感，又彰显出其自身的个性。角部的270°超规格弧形玻璃刻画出塔楼的细节，也为酒店及办公人群提供了独特的视野感受。低层商业街区的建筑外立面采用了现代中式语言风格，横向大尺度的开窗呼应其商业用途，古典中透出时尚的气息。

天目里
OōEli

项目地点：浙江省杭州市
建筑面积：230,000m²
设计/竣工：2012/2020
Location: Hangzhou, Zhejiang
Floor Area: 230,000m²
Design/Completion: 2012/2020

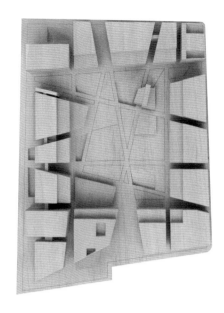

OōEli is a joint development between the fashion company JNBY and the architectural firm GOA, and the renowned Renzo Piano Building Workshop (RPBW) was entrusted with leading planning and design collaboration. In this remarkable undertaking, GOA assumed the unique position of being both the executive designer and the client.

OōEli integrates two corporate headquarters, an art museum, an art center, a boutique hotel, offices, show fields, and art commerce, among other features. Its design aims to establish an oasis-like "city parlor" with buildings along the periphery encompassing a 130m × 95m central plaza.

Diagonal pathways divide architectures into pleasing volumetric proportions. These pathways connect the plaza and the city, enabling people, breeze, and sunlight to enter from all directions. The setback terraces facing the center enhance the spatial scale, while green landscapes extending from underground to the terraces and two water mirrors at the park's center bring a touch of nature throughout the seasons. These elements create a soft green core that reflects Hangzhou's natural landscape.

Transfer elevators located at the four corners of the site serve as the sole pathway from the underground garage to the ground level and guides pedestrians to traverse the courtyard, making the plaza the true "core." Whether individuals are heading to work, exploring museums, attending shows, relaxing, or dining out, they all inspire spatial vitality.

天目里由时尚品牌江南布衣联合goa大象设计共同出资建设，并邀请伦佐·皮亚诺先生领衔规划设计，大象设计同时担任执行建筑师和业主的双重角色。

项目包含办公楼、美术馆、艺术中心、秀场、设计酒店及艺术商业等功能，是一个综合性的艺术园区。设计的核心想法是将建筑贴合基地的外沿布置，围合出长130m、宽95m的中央广场，以此塑造一个绿洲般的"城市客厅"。

几条斜向路径将建筑体量分割为适宜的尺度，也创造出广场和城市的联系通道，来自四面八方的人流与空气得以在广场流通；各建筑体量面向广场局部退台，进一步优化了广场的空间尺度；从广场延伸至楼宇露台的绿色植物、位于广场核心的水面则带来四时变幻的自然情境，与杭州的城市特质深度耦合，是整座园区的温柔内核。

从地下室停车区通往地面的必经路径是分散在广场四个角落的转换电梯，由此，由地下及地上交通抵达的行人都须经过或绕行广场，才可到达要去的塔楼——这让广场成为真正的"核心"。日常办公的人、去美术馆看展览的人、去秀场观看时装秀的人、在广场上闲坐的人、在咖啡厅和餐厅吃饭的人，都成为"城市客厅"的使用者。

For

Working

办公 & 产业

浙商银行总部
China Zheshang Bank Headquarters

宇视科技总部
Uniview Headquarters

西子智慧产业园
Xizi Wisdom Industrial Park

杭州东站花园国际
Hangzhou East Railway Station Garden International

海口五源河创新产业中心
Haikou Wuyuanhe Intelligent Creative Collective

石家庄中央商务区办公楼
Shijiazhuang CBD Office Towers

宁波智造港芯创园
Ningbo Intelligent Manufacturing Port

江南布衣仓储园区
JNBY Warehousing Logistics Park

台州数字科技园
Taizhou Digital Technology Park

浙商银行总部
China Zheshang Bank Headquarters

项目地点：浙江省杭州市
建筑面积：310,000m²
设计/竣工：2018/—

Location: Hangzhou, Zhejiang
Floor Area: 310,000m²
Design/Completion: 2018/-

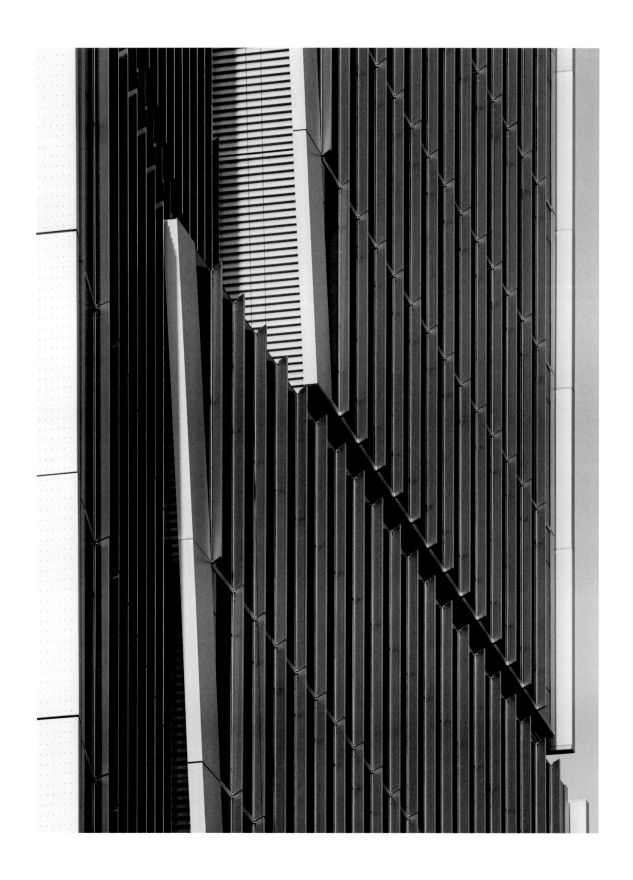

In the core of Qianjiang Century City, the China Zheshang Bank Headquarters exudes a commanding presence on the southern bank of the Qiantang River as a striking manifestation of the corporate spirit and values.

The headquarters encompasses a nearly 250m-tall office module, service modules ranging from 24m to 60m in height, and a three-story underground module for basic supports. The design incorporates two crucial axes intersecting at the entrance courtyard to foster integration between the office and service modules. The vertical "work axis" is the primary vertical circulation of the 136,000m² office tower, and the horizontal "relax axis" features a spacious atrium connecting various functions such as conferences, training, dining, fitness, entertainment, medical facilities, and lodging.

The design incorporates a stacking of the traditional measurement tool "Dou" as a motif, reflecting the identity of the financial enterprise and evoking a sense of terracing and ascending. This idea is integrated into various design aspects, including master planning, podium design, spatial layout, landscaping, and roof details, resulting in a cohesive architectural style. The north and south slicing gestures produce a dynamic Z-shaped facade effect. Considering CZBANK's regional background, the design incorporates the imagery of traditional Zhejiang sloping roofs into various architectural elements, such as the entrance canopy, the podium's horizontal sunshades, and the tower's vertical louvers. The landscape system features greenery at different elevations, providing workers with a "garden office" experience.

A prefabricated construction approach for the curtain wall system has been employed to promote sustainability. The curtain wall uses double-layer insulating glass, combined with the facade sunshade system to significantly reduce daily energy consumption. Upon completion of the project, a new "Green Financial Landmark" will be established in the financial district along the Qiantang River.

浙商银行总部位于杭州钱江世纪城核心区，建筑将昭示浙商银行的企业精神与价值观，同时也将成为钱塘江南岸天际线的重要组成部分。

该总部由近250m高的主体办公模块、24m至60m高的办公服务模块以及地下三层的基础保障模块所组成。高效工作"芯"是塔楼的垂直交通动脉，贯穿136,000m²的塔楼办公区。健康生活"芯"则以长中庭的形式连接会议、培训、餐饮、健身、娱乐、医疗服务、住宿等配套功能。两条动脉交汇于建筑入口庭院，实现了办公模块与服务模块的整合。

设计以传统计量工具"斗"为母题，藏寓金融企业交易汇兑的业务特征，层叠的"斗"展现出节节退台、向上生长的建筑形象。南北两侧的削切手法带来具有动感的Z字形立面效果。"斗"的母题也被应用至总图规划、裙房形体、主要空间轮廓、景观肌理、屋顶细部等不同维度的设计中，使建筑整体得到统一的表达。基于浙商银行的地域基因，建筑师撷取具有浙江传统建筑意味的坡屋顶元素，经由抽象处理，化用于入口雨棚、裙房遮阳板、塔楼遮阳构件等不同部位。绿色植物被引入不同标高的室外空间，为员工带来"花园办公"的日常体验。

建筑外围护系统的设计及建造皆贯彻了可持续性理念。幕墙采用双层中空玻璃，结合外立面遮阳体系大幅降低日常能耗。项目竣工之日，钱塘江畔的金融聚集区将再添一座崭新的"绿色金融地标"。

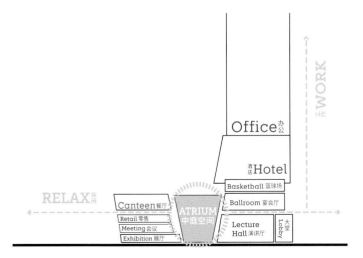

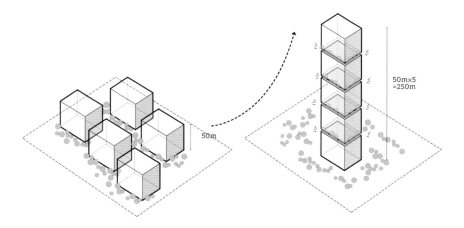

宇视科技总部
Uniview Headquarters

项目地点：浙江省杭州市
建筑面积：136,700m²
设计 / 竣工：2017/2023
Location: Hangzhou, Zhejiang
Floor Area: 136,700m²
Design/Completion: 2017/2023

The Uniview Headquarters is located in the Binjiang IoT Industrial Park, one of Hangzhou's key high-tech industry hubs. This headquarters is an advanced intelligent industrial park integrating offices, research and development, conferences, and reception facilities.

The site's western boundary connects to the southern end of an urban green axis. Along this axis, the design reveals a 100m-tall tower, a horizontally extended R&D center, and a spacious podium. The interior shared spaces are modularly designed to foster collaboration while maintaining the hierarchical confidentiality of the organization, thereby establishing a technology hub of innovation, dynamism, and creativity.

On the ground floor, a 120m-long and 7m-wide elevated corridor, the West Entrance, serves as the pedestrian artery for the headquarters. It connects the external park, sunken courtyard, and entrances to the R&D area, facilitating easy navigation and efficient mobility between the podium and working areas.

The underground restaurant accommodates 1,200 people. It serves as a leisure hub with a gym, library, café, and convenience stores. Adjacent to the restaurant, the sunken plaza and atrium invite ample natural light access and provide direct access to different areas.

The R&D center offers a 5,000m² flat floor space, providing optimal flexibility for organizational needs. The unique crescent-shaped atrium maximizes the influx of natural light and provides lovely views, while the Z-shaped staircase winding around it facilitates vertical traffic flow to encourage collaboration. Spanning across the second and third floors, the auditorium accommodates 400 seats and seamlessly integrates with the park landscape on its west. The wide staircase ensures a connection with the south hall, creating a venue suitable for large events or training sessions.

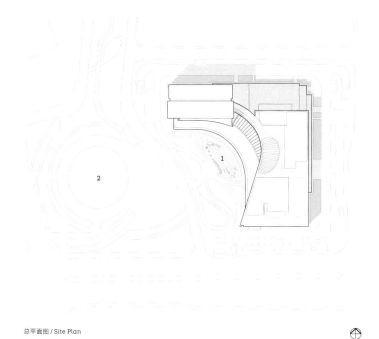

总平面图 / Site Plan

1　下沉广场　　1　Sunken Square
2　城市公园　　2　City Park

宇视科技总部位于杭州高新产业集中的滨江区物联网产业园内，是融合办公、研发、会议和接待于一体的新型智能园区。

基地西侧衔接片区绿化主轴的南端。设计顺应绿化轴线，主体量由100m高的塔楼、水平展开的研发中心及超级基座组合而成。多样化共享空间置入模块化的布局，可以在解决企业内合作、分级保密性问题的同时，赋予企业总部作为科技企业的开放、年轻气质。

长120m、宽7m的一层架空弧形走廊是总部的西入口，也是主要的人行动脉。它衔接着外部公园、地下庭院及研发区域的出入口，引导不同需求的人们进出建筑的不同部位，极大程度上提升了超级基座与办公区域之间的通行效率。

位于地下一层的餐厅可同时容纳1,200人就餐，餐厅连同健身房、图书馆、咖啡馆及便利型商业，构成大楼的生活底盘。餐厅外部的下沉广场及带状中庭既满足了空间的采光需要，又便于人群的快捷流通。

研发中心以5,000m²的单层面积提供最大化的空间灵活性。月牙形的共享中庭将阳光与景观引入，围绕中庭的Z字形楼梯增强了低区各层的上下往来，也塑造了研发中心鼓励交流与碰撞的工作氛围。跨越二、三层的400人规模的阶梯教室同样引入了西侧的公园景观，宽阔的开放式楼梯保证了其与南大厅的联系，也是企业内大型活动及培训的场所。

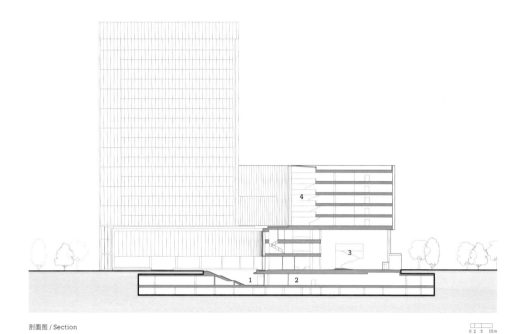

剖面图 / Section

1	下沉广场	1	Sunken Square
2	员工餐厅	2	Staff Canteen
3	大堂	3	Lobby
4	研发中心中庭	4	R&D Center Atrium

西子智慧产业园
Xizi Wisdom Industrial Park

项目地点：浙江省杭州市
建筑面积：310,000m²
设计/竣工：2018/2023

Location: Hangzhou, Zhejiang
Floor Area: 310,000m²
Design/Completion: 2018/2023

In 2008, Hangzhou Boiler Factory, a prominent industrial cluster in the city's northern region, relocated from its original site to Dingqiao, which was then a suburb. As a result of rapid urbanization, the former-farmlands and industrial zones in Dingqiao gradually gave way to residential areas. Since 2015, the Hangzhou Boiler Factory in Dingqiao and the surrounding 215,000m² area have embarked on three phases of new development plans. The objective is to establish a modern industrial community that integrates industrial offices, residential areas, and recreational facilities, achieving a spatial and industrial evolution.

The design incorporates two strategies to optimize land utilization: firstly, by integrating commercial formats and essential supporting facilities within the existing factory buildings, maximizing land value and benefiting the surrounding area; and secondly, by diversifying functions, such as cafeteria, bookstore and cultural programs, and connecting public spaces with enchanting gardens, thus enhancing the vitality and inclusiveness of the community.

The Phase 1 office park features an intertwining of building clusters and gardens. The ground circulation system separates vehicles and pedestrians. It offers diverse mobility experiences with 15m-wide urban roads, 4m-wide internal roads, and 2m-wide landscape footpaths. Additionally, the park incorporates covered corridors and canopies to provide shaded areas where people can pause and relax.

The Phase 2 factory renovation preserves an industrial reminiscence by retaining the existing steel structure and column-span. A new functional layout features a 20,000m² Sam's Club, a 30,000m² shopping center, and a 27,000m² parking garage. The original steel covering is transformed into a striking reddish-brown metal roof with a zigzag skylight system that utilizes a 12m module. The folding effect on the facade generates a rhythmic interplay of empty and full spaces, breaking up the monotony of the street interface.

Formerly a packaging plant and outdoor storage, the Phase 3 site is now an advanced campus for high-tech enterprises' production, research, and development. The design of the building facade combines elements such as glass, aluminum panels, and horizontal grilles, breaking away from the conventional perception of cold and uninviting industrial structures.

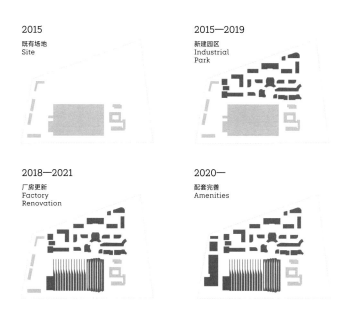

自2008年起，杭州城北工业集群的代表杭州锅炉厂（杭锅）外迁至彼时尚作为郊区的丁桥一带。随着城市的迅速发展，丁桥的农田及厂区逐渐被成规模的居住区取代。自2015年起，杭锅丁桥厂房及其周边约215,000m²的地块分三期启动新的发展计划，旨在建设成为产业办公与生活休闲于一体的新时代产业社区，以实现空间及产业双重意义上的转型。

面对这片存量工业用地，建筑师提出两点设计策略：一是在旧厂房中置入商业及基础配套，以激发土地价值和服务周边区域；二是在产业园区中置入食堂、书店、音乐厅等丰富业态，通过花园密布的公共空间串联起具有活力的综合型社区。

一期地块的办公园区中，建筑与花园相互交错。人车分流的地面上，15m宽的街道、4m宽的园区主路及2m宽的花园小径交织出丰富的步行层次。建筑二层以下以连廊和雨棚组成连续灰空间，满足人们停驻的需求。

二期地块的厂房更新方案保留了既有的钢结构及柱网作为记忆的延续，同时置入约20,000m²的山姆会员店、30,000m²的集中式商业及27,000m²的停车楼。原有的彩钢板屋面被更新为红褐色金属屋面，以12m为模数的锯齿状天窗体系为下方大空间引入自然光；立面上的折板造型形成虚实韵律，一改既有厂房沿街面单调、冗长的面貌。

三期地块原为杭锅包装车间和露天堆场，改造后成为高新技术企业的集生产、研发于一体的厂区。厂房立面综合运用玻璃、铝板、水平格栅材料，颠覆了工业建筑冰冷而缺乏生气的固有形象。

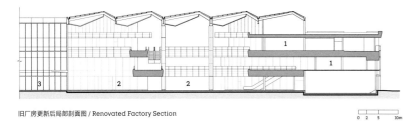

旧厂房更新后局部剖面图 / Renovated Factory Section

1	商铺	1	Retail
2	商业中庭	2	Commercial Atrium
3	消防通道	3	Fire Exit

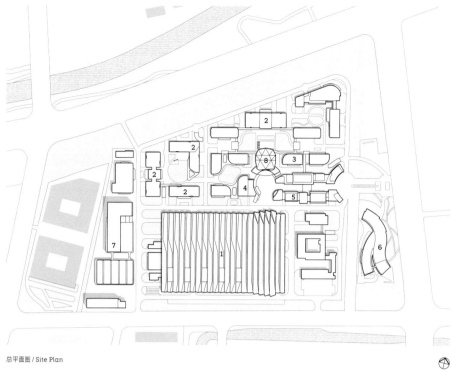

总平面图 / Site Plan

1	集中商业	1	Centralized Commercial
2	智能制造中心	2	Intelligent Manufacturing Center
3	研发中心	3	R&D Center
4	总部研发	4	Headquarters R&D
5	创意办公	5	Creative Office
6	办公	6	Office
7	中试车间	7	Pilot Workshop
8	会议中心	8	Conference Center

杭州东站花园国际
Hangzhou East Railway Station Garden International

项目地点：浙江省杭州市
建筑面积：138,500m²
设计/竣工：2017/2022

Location: Hangzhou, Zhejiang
Floor Area: 138,500m²
Design/Completion: 2017/2022

Hangzhou East Railway Station is one of Asia's largest railway hubs and a crucial node within China's "Eight Vertical and Eight Horizontal" high-speed railway system. Situated at a pivotal intersection along the east axis of this transit hub, Garden International office park is within a 1km walking distance from the railway station. Its design objective is to create a vibrant core that connects urban vitality dots and facilitates regional business formats. The office park comprises three garden-style office buildings, a star-rated hotel, and a commercial block. This functional integration contributes to developing a distinct "closed-loop system of commercial and business."

A 37° angle exists between the site and the north-south axis, and the buildings are arranged along the city roads to optimize floor-area and built-to-line ratios. Through a "twist" gesture, the design imparts a bending form to the rectangular volumes. Four curved lines create an inclusive yet unobstructed "pinwheel" layout, where the buildings embrace and echo each other while maintaining independence. The winding spaces between the buildings resemble a "natural canyon." Beyond meeting sunlight, ventilation, and setbacks requirements, this distinct master plan strategy respects the integrity of the urban interface while expressing its architectural personality.

The facade combines aesthetics with practicality. Horizontal aluminum components are the primary elements, aligning with the building's curves and allowing for the discreet placement of openable windows. A dynamic three-dimensional "twist" of the protruding 1.2m aluminum components mitigates direct sun exposure from the west while ensuring abundant natural light penetration from the south. A cost-effective window wall system is a high-quality alternative to traditional curtain wall systems.

杭州东站是中国高铁网"八横八纵"的关键节点之一，也是"亚洲最大的铁路交通枢纽"之一。东站花园国际位于东站枢纽东轴线一侧的重要路口，距离杭州东站步行距离不足1km。本项目旨在成为串联城市活力环的活力点，为区域商务业态赋能。项目包含三幢花园式办公楼、一幢星级酒店和漫步式商业街区。多元的业态循环互动，形成独特的商业商务生态链。

基地与正南北向存在37°的夹角，建筑沿城市道路布置以提高建筑容量及贴线率。通过一个巧妙的"扭转"动作，设计赋予矩形体量一种弯折的形态。四条弯折线条构成"围而不堵"的风车状布局。成环抱之势的楼宇相互呼应又各自独立，曲折灵动的内部空间如同一座自然峡谷。这一独特的总图策略不仅满足日照、通风、退界等条件，更在尊重城市界面完整性的同时实现了建筑的个性表达。

建筑立面巧妙地融合了形式感与功能性。考虑到弯弧的体量和隐藏开启扇的需求，建筑师采用横向铝板构件作为立面主要元素。为满足不同立面朝向的差异性需求——西侧趋避西晒、南侧引入日照，建筑师对出挑1.2m的铝板构件进行了三维扭转处理。此外，高性价比的窗墙体系实现了近似于幕墙体系的品质感。

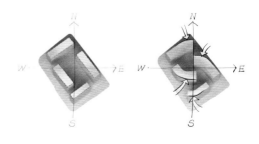   

体量 / Volume　　　矫正 / Correction　　　　　　　　环抱 / Encircle　　　流动 / Circulation

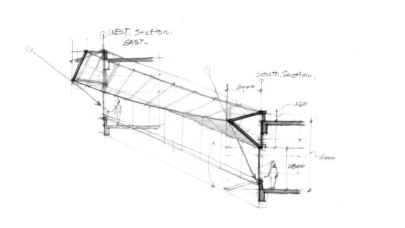

海口五源河创新产业中心
Haikou Wuyuanhe Intelligent Creative Collective

项目地点：海南省海口市
建筑面积：415,000m²
设计/竣工：2021/—

Location: Haikou, Hainan
Floor Area: 415,000m²
Design/Completion: 2021/-

Haikou Wuyuanhe Intelligent Creative Collective (ICC) is a science and technology industrial cluster located adjacent to Wuyuanhe National Wetland Park. With a core emphasis on ecological priority, the design centers around climate adaptation, public integration, and compound greening. Following the Park-oriented Development (POD) concept, it seeks to maximize the preservation of the original wetland park and incorporate healthcare, recreation, social, and business facilities within an eco-friendly framework.

The master plan establishes a north-south green axis and an east-west vitality axis to connect different plots within the site. These axes divide the park into four main zones: headquarters offices, industrial services, research and design, and innovative finance, creating a circular economy industrial chain encompassing service, incubation, research and development, production, and exhibition. The open cluster-based structure, human-scale design, and diverse spatial combinations facilitate the fusion of commercial, cultural, recreational, and supporting services, establishing a dynamic and convenient 5-minute service circle. With its all-day functionality, the park becomes an urban parlor embraced by nature.

The open green spaces and slow circulation networks between clusters generate a dynamic and accessible ecological landscape system. Integrating waterfront architecture with the surrounding natural resources forms an inviting and vibrant water edge. Thus, the project reveals a vast, boundaryless park.

The design integrates sustainable development strategies. The architectural facades address Haikou's tropical climate by balancing sun shading and scenic views demands. The ground level incorporates skylights and lush greeneries, enabling natural light penetration and improved ventilation. This creates a more comfortable office environment and contributes to the concept of a sponge city.

本项目毗邻海口五源河国家湿地公园，是一片坐落于优美自然之中的科技产业群落。在生态优先的核心原则之下，设计围绕气候适应、聚合共享、复合绿化三大思路展开。基于POD开发理念，设计最大程度保留场地内的原生湿地公园，将健康、娱乐、社交、商业等公共服务功能组织于景观基底之上。

总图规划利用纵贯南北的生态绿轴和横跨东西的活力主轴串联起各地块。这两条轴线将园区划分为总部办公、产业服务中心、设计研发中心、金融聚集区四大分区，形成服务—孵化—研发—生产—展示的循环经济产业链。开放的组团式结构、人性化的尺度设计、多元化的空间组合推动着商业、文化、娱乐、配套服务的有机互融，帮助实现5分钟服务圈的构建。全功能、全时段的功能组合让园区成为一个与森林和湿地相拥的都市客厅。

组团间的共享绿地与慢行交通脉络结合两大主轴，构建活力、开放的生态景观系统。滨水建筑与自然资源深度融合，勾勒出亲切友好的公共水岸。由此，园区如同一个大型的无边界公园。

设计于诸多层面融入可持续发展理念：建筑立面的设计回应了当地气候条件，平衡了遮阳与观景需求；地下空间通过设置采光井和景观绿化引入自然光线、增强通风效率，在提升办公场所舒适性的同时也回应了海绵城市的建设纲领。

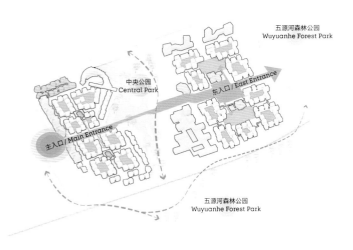

石家庄中央商务区办公楼
Shijiazhuang CBD Office Towers

项目地点：河北省石家庄市
建筑面积：136,800m²
设计/竣工：2019/—

Location: Shijiazhuang, Hebei
Floor Area: 136,800m²
Design/Completion: 2019/-

The project consists of three office towers, arranged in a north-south orientation, serving as the northern hub of the upcoming Shijiazhuang Central Business District. The site is bordered by urban green belts on the east and west and a leisure park to the south. Furthermore, several historic buildings add to the area's cultural significance within a one-kilometer radius to the south.

The design aims to create a well-balanced urban living circle in the future Central Business District, focusing on efficiency, ecology, and vitality. The master plan treats the three parcels as an integrated entity, adhering to the urban design principle of narrow streets and dense roads. The circulation of vehicles is optimized to maximize the availability of pedestrian spaces.

The three towers ascending from the southern to the northern depict a dynamic skyline and act as a prominent entrance to the district. The design eliminates the claustrophobic feeling typically associated with tall buildings by adopting dividing, segmenting, and twisting gestures. The north-south axis divides each tower into two thin volumes, while the vertical segmentation and horizontal twisting enhance its dynamic stance. The sweeping curves add a unified and symbolic touch to the architecture cluster, harmonize with the extending green spaces, and reflect the arched features in historic structures like the Shijiazhuang Zhengtai Hotel and Dashiqiao.

The serrated facade profile delivers a distinctive visual effect. When viewed from one direction, The anodized perforated aluminum panels are concealed behind large glass surfaces, delivering a sleek and transparent aesthetic. From the other direction, these two materials generate a rhythmic facade texture through consistent modular control.

本项目由南北向分布的三座塔楼组成，是未来石家庄中央商务区北端的金融商务高地。场地东西两侧均为连续的城市绿带，南侧为休闲公园，向南1km范围内分布着石家庄的重要历史建筑。

项目作为未来中央商务区的一部分，其设计目标为实现都市生活圈效率、生态、活力的平衡共生。总图规划将三个地块综合考虑，遵循窄街密路的城市设计原则，通过车行流线的集约化处理，提供充裕的步行空间。

三座塔楼自南向北渐次高起，建筑高度的变化描绘出动态而立体的天际线，塑造出中央商务区的北门户形象。为化解高密度建筑的压迫感，设计对三座塔楼进行切分、裁剪、扭转。沿南北轴线的切分赋予塔楼纤薄感，垂直分段裁剪及水平扭转则为建筑带来优雅的动势。大尺度弧线成为建筑造型的母题，弧线的组合强化了建筑集群的整体感，既契合场地双侧绿带的延伸趋势，也与石家庄正太饭店、大石桥等历史建筑中的拱形元素遥相呼应。

建筑幕墙的锯齿状截面呈现独特的视觉效果。从一侧观察，阳极氧化穿孔铝板被隐藏于大面积玻璃面之后，立面显得简洁通透；而由另一侧看，玻璃面与穿孔铝板在统一模数控制下构成有韵律的界面。

 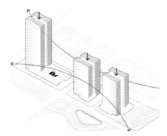  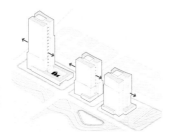

宁波智造港芯创园
Ningbo Intelligent Manufacturing Port

项目地点：浙江省宁波市
建筑面积：310,000m²
设计 / 竣工：2020/—
Location: Ningbo, Zhejiang
Floor Area: 310,000m²
Design/Completion: 2020/-

Ningbo Intelligent Manufacturing Port is the city's first industrial park dedicated to intelligent manufacturing, living, and operations. It is a key initiative in Ningbo National Hi-Tech Industrial Development Zone's future vision. The project contains intelligent factories, research and development centers, a central park, a maker community, rental apartments, commercial facilities, and other supporting services, aiming to establish a comprehensive, innovative platform that integrates industry, city, people, and culture in synergy.

The master plan is designed around the "one ring, one axis, and multiple gardens" concept, with an ecological landscape axis traversing the northern and southern parcels. Elevated walkways, sky corridors, and public terraces connect various building clusters, creating a vibrant and dynamic environment that encourages communication and sharing.

The project comprises multi-story and high-rise industrial buildings, utilizing stacking gestures and variable modules that respond to the characteristics of a next-generation industrial park. The vertical stacking of production spaces improves land utilization efficiency, while the modular combination accommodates manufacturing, pilot testing, research, and office work. The 7m-high ground floor caters to large production and exhibition space requirements.

The architectural design draws inspiration from the dynamic rhythm of "waves" and the interconnectedness of "computer chips." Streamlined planar contours, continuous horizontal strip windows, and gradually protruding curved aluminum panels create a unified, futuristic appearance that celebrates the concept of "technology." The architectural cluster features a diverse color palette, with perforated copper aluminum sheets injecting a touch of vibrancy to the primary tone of light grey aluminum panels and the light blue-grey glass, enhancing the overall vigor.

本项目是宁波首个以智能制造、智慧生活和智慧运营为核心的工业产业园，也是宁波国家高新区未来计划中的重要发展平台。项目汇集智能厂房、研发中心、中轴公园、创客社区、人才公寓、商业配套等，旨在打造"产、城、人、文"一体的复合型创新平台。

一条生态景观轴贯穿园区的南北地块，在"一环一轴多园"的格局之下，各建筑组团间通过底层架空、空中步道、共享露台等互相渗透、连接，以此塑造园区内交流共享的氛围。

园区包括多层及高层产业建筑，采用空间叠加、可变模块两项策略回应新一代产业园区的特质。生产空间的垂直叠加提升了土地使用效率，模块化单元的灵活拆分组合则适应不同企业生产、中试、研发及办公的需求。建筑首层挑高7m，可满足大空间生产及展示的需要。

建筑造型以"水波"的动态韵律、"芯片"的链接和交错为灵感，利用流线型的平面轮廓、连续环通的水平条窗、渐变出挑的弧形铝板等设计元素打造富有"科技感"与"未来感"的整体形象。建筑群落色彩丰富，在浅灰色铝板及浅蓝灰色玻璃所构成的主基调之上，铜色穿孔铝板作为亮色穿插其中，为园区增添一丝活力。

1	生态景观轴	1	Ecological Landscape Axis
2	高层厂房	2	High-rise Plant
3	多层厂房	3	Low-rise Plant
4	办公	4	Office
5	酒店式公寓	5	Serviced Apartment
6	商业	6	Commercial

总平面图 / Site Plan

104　办公&产业　　For Working

江南布衣仓储园区
JNBY Warehousing Logistics Park

所在地址：浙江省杭州市
建筑面积：91,000m²
设计 / 竣工：2016/2019
Location: Hangzhou, Zhejiang
Floor Area: 91,000m²
Design/Completion: 2016/2019

The JNBY Warehousing Logistics Park is located within the Xiaoshan Economic and Technological Development Zone, serving as a hub for manufacturing, sorting operations, and logistics. Adjacent to the former JNBY manufacturing base to the south, with the Hangzhou-Ningbo Expressway to the north and the urban elevated highway to the west, the project aims to create a well-organized spatial layout and visually appealing factory park at this significant intersection.

The main entrance is positioned in the southeast corner to increase production and logistics efficiency and mirror the former production factory's main entrance across the road. The production and logistics workshops are symmetrically arranged in the east and west sections. The truck loading area is positioned to the north, ensuring smooth logistics circulation while maintaining a clean street-front interface in the south. Furthermore, the external circulation core provides the building with a more integral logistics and storage space internally.

The architecture celebrates JNBY products' softness, crispness, purity, and simplicity. Its facade features corrugated metal panels, enveloping the expansive interior spaces and external circulation core, resulting in an integration of form and function. The north interface emphasizes architectural legibility by incorporating an extending, uninterrupted arc. The building's sculptural volume resonates with the garment tailoring while the curves and folds on the facade reflect the tactile softness of fabrics, with a fine-grained texture creating a delicate interplay of light and shadow.

The lower level of the building showcases a continuous glass curtain wall, while the upper level incorporates scattered square windows, introducing visual interest and rhythm to the facade. This design gesture provides ample natural light for the ground-floor workspace and ensures compliance with fire safety regulations on the upper level. Additionally, windows oriented towards the atrium enhance internal lighting and ventilation.

本项目位于杭州萧山经济技术开发区，是集生产、拣选作业、物流为一体的生产物流基地。项目基地与南侧的江南布衣老生产基地一路之隔，北侧即为杭甬高速，西侧为城市高架路。如何在干道交汇处塑造一座布局合理、形象富有魅力的厂房园区是本项目研究的课题。

为了提高生产和物流的效率，园区主入口布置于地块东南角，与马路对面的老生产基地物流主入口相对应。生产物流厂房分为对称的东、西两部分，货车装卸货区位于北侧，确保便捷的物流流线及整洁的南侧临街界面。厂房的交通核外挂，为建筑内部提供更完整的物流储藏空间。

建筑设计延续江南布衣产品柔软、有型、纯净、朴素的特质。其立面利用金属波纹板可成弧的特性，将大空间与外挂交通核包裹，实现表皮与功能的统一。面向高速公路的北立面以完整的弧面最大限度展现建筑的标志性。立面上重叠的弧面与褶皱隐喻着布料的柔软质感；建筑体量的雕塑感则使人联想到时装的版型；波纹钢板表面肌理在弧面中产生细腻的光影过渡。

建筑底层连续的玻璃幕墙以及上部不规则的方窗不仅为立面节奏带来变化，还满足了底层作业区采光以及上层消防救援等需求。与此同时，建筑面向中庭的开窗优化了内部的采光通风。

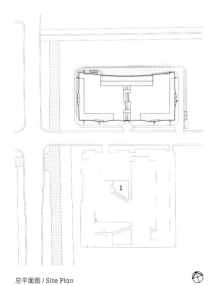

1　既有生产基地

1　JNBY Manufacturing Base

总平面图 / Site Plan

台州数字科技园
Taizhou Digital Technology Park

项目地点：浙江省台州市
建筑面积：231,450m²
设计 / 竣工：2022/—

Location: Taizhou, Zhejiang
Floor Area: 231,450m²
Design/Completion: 2022/-

108　办公&产业　　For Working

The Taizhou Digital Technology Park is Taizhou's first Innovative Industrial Zoning (M0) District. Its primary objective is to develop a platform for incubating digital technologies, R&D in manufacturing, innovation services, and digital industries.

The master plan draws inspiration from the harmonious coexistence of mountains, sea, and cityscape, reflecting the urban pattern seen in Chinese landscape paintings. It utilizes a spatial layout that creates dispersed and dynamic clusters where blank spaces and scenic views interact and intertwine.

Integrating industrial and living spaces provides a comprehensive solution that caters to modern enterprises. The industrial sector offers adaptable spaces for companies at various stages of development, enabling customization based on specific requirements. The available spaces can be subdivided into multiple cells, ranging from 500-3,000m², and arranged vertically or horizontally to accommodate different layouts. Alongside the industrial areas, the public spatial structure enhances the park's natural beauty and encourages social engagement. The converging diagonal pathways in the central garden attract visitors from inside and outside the park. Moreover, the exhibition area, located in the site's center and facing the main road, becomes an impressive technological display.

The design incorporates sustainable strategies to achieve net-zero energy consumption. Urban ecological wind corridors are established between buildings, and their spacing is carefully controlled through sunlight analysis. The lower office area on the south side features tall atriums that provide abundant natural light and fresh air. These atriums are adorned with vertical greenery and interactive corners, creating pleasant spatial encounters. Sunshades have been installed on the upper facade sections to reduce glare from the west. Additionally, the park includes two exhibition halls with a near-zero energy consumption theme, promoting a green future vision.

台州数字科技园是台州首个M0用地科技创新园区，旨在成为数字核心技术、数字智造研发、数字创新服务、数字内容创新产业的孵化平台。

受台州"山海水城、城景相依"的城市格局启发，设计将中国山水画的空间布局手法与产业园区相融合，以留白、对景手法为建筑群落带来错落有致、处处透景的灵动感。

设计将产业空间与生活空间有机结合，以充分适应创业发展的需求。产业空间由"产业细胞"构成，"产业细胞"的面积划分可实现500m²至3,000m²的弹性分配，匹配不同发展阶段企业的需求。在水平向组合方式之外，三层垂直叠加的组合模式为企业提供更为个性化的布局可能。在产业空间之外，设计还刻画了承载绿色自然与人群交往的场所，使之成为园区的骨骼。中央花园之上，通往四面八方的斜切路径将园区内外的人群汇聚。科技秀场位于场地中央，面向主干道拥有颇具气势的展示面。

面向净零能耗的目标，园区整体设计遵循了碳中和原则。楼栋之间规划有城市生态风廊，楼栋间距皆以精细化的日照分析为控制依据。南侧的低区办公设有通高中庭，保证内部采光与通风，其两侧设置立体绿化及交流空间，为使用者提供丰富的空间体验；高区山墙面设置有水平遮阳板，用于减少太阳西晒带来的眩光。园区内特别设有两处近零能耗示范区展厅，向访客传达有关绿色未来的美好目标。

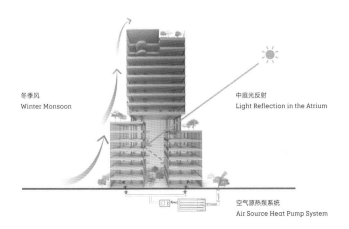
冬季风 / Winter Monsoon
中庭光反射 / Light Reflection in the Atrium
空气源热泵系统 / Air Source Heat Pump System

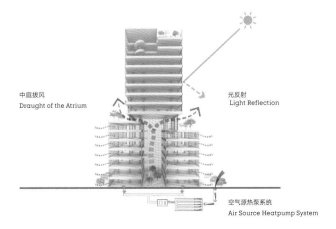
中庭拔风 / Draught of the Atrium
光反射 / Light Reflection
空气源热泵系统 / Air Source Heatpump System

For

Urban Renovation

城市更新

祥符桥传统风貌街区
Xiangfu Bridge Historic District Renewal

上海北外滩32街坊更新
Shanghai North Bund Neighborhood 32 Renewal

弘安里
Hong'anli

中海顺昌玖里
China Overseas Arbour

里直街
LiZhi Street

祥符桥传统风貌街区
Xiangfu Bridge Historic District Renewal

项目地点：浙江省杭州市
建筑面积：66,100m²
设计／竣工：2019/2023
Location: Hangzhou, Zhejiang
Floor Area: 66,100m²
Design/Completion: 2019/2023

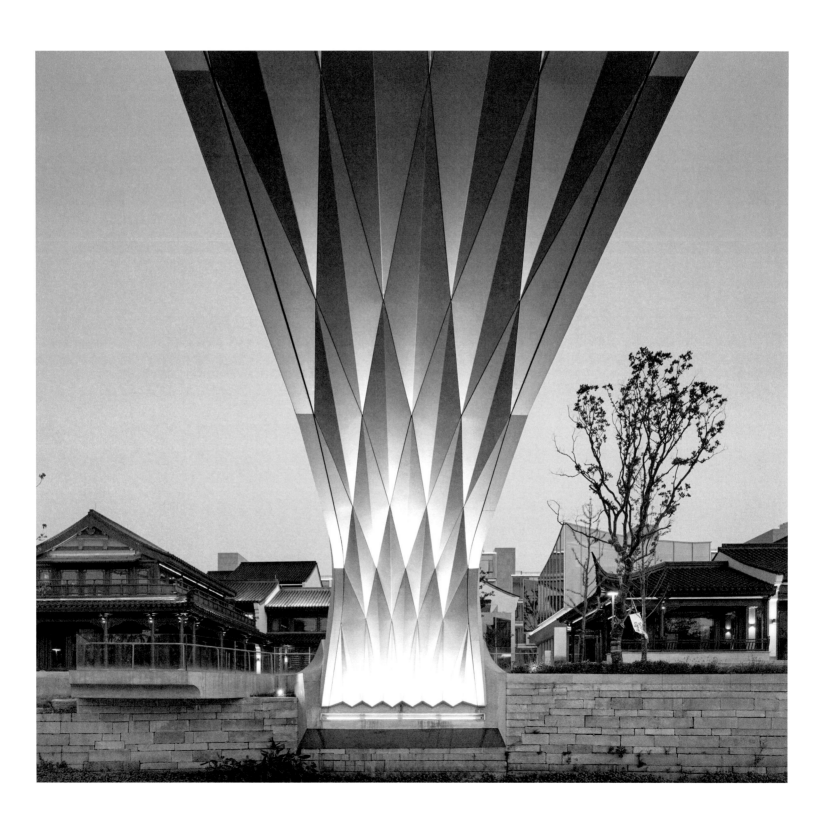

In recent years, the Xiangfu Bridge Historic District Renewal has been one of Hangzhou's most remarkable urban revitalization projects. It embodies the rich heritage of the Song Dynasty's canal culture, epitomized by the nearly 800-year-old Xiangfu Bridge, a national-level cultural relic. In modern history, the Xiangfu Bridge District has emerged as an industrial district connecting the city to the suburbs, where residents embrace a vibrant waterfront lifestyle. Existing structures line the Wulitang River in the downtown area, while the old streets and alleys exude a bustling sense of life. However, the sustainable development potential of this congested neighborhood is limited.

Thus, the design scheme aims to revitalize the region while preserving its distinct local "marketplace" character by enabling a fusion of historical memories and modern lifestyles to provide high-quality public spaces for the surrounding cities and to attract new life to the old streets.

It introduces the concept of integration in "spaces," "businesses," and "histories," emphasizing the neighborhood's flexible and interconnected spatial layout and the constant presence of commercial and cultural activities. By conducting a typological analysis of the traditional "marketplaces," the various dimensions of "courts," "yards," "streets," "alleys," "terraces," and "gardens" are organized as distinctive spaces within the neighborhood that serve as a collection of traditional memories. The master plan leverages the Wulitang River to establish a dynamic waterfront area extending into the cultural and historic streets on the north and south riverbanks. Two newly constructed pedestrian bridges, strategically positioned over 70 meters from the Xiangfu Bridge, facilitate connectivity between the riverbanks while preserving the Xiangfu Bridge's scenic view. In response to the community's functional needs, a new school and senior community are incorporated with open-ground pathways that integrate seamlessly with the neighborhood to promote resource sharing.

The renewal of Xiangfu Bridge Historic District was conducted using the EPC general contract model. This approach represents a significant endeavor in large-scale urban renewal projects where the design firm assumes the project management role.

祥符桥传统风貌街区是杭州近年来最重要的城市更新项目之一。其作为中国宋代运河文化的传承地，街区中的全国重点文物保护单位祥符桥的历史可追溯到约800年前。在近现代历史中，祥符桥街区成为老工业区及城郊结合部，居民依水而居，原始建筑沿五里塘河形成一字排开的街市格局。老街小巷积淀着浓浓的生活气息，但街区整体十分拥塞，在空间上缺乏持续发展的潜力。

更新计划在保留街区"市井"风貌的基础上进行活力激发，历史的记忆与现代的生活方式将在此融合，街区将为周边城市提供高品质的公共空间；同时，外来访客也将为古老的街区注入新的活力。

设计提出"空间融合、业态融合、时间融合"的概念，强调街区空间内商业与文化活动的灵活性、可渗透性以及全时性。通过对传统"市井"空间的类型学分析，多种尺度的"庭""院""街""弄""台""园"作为特色空间被组织在街区中，成为传统记忆的载体。更新总图依托五里塘河打造滨水活力带，向南北两侧文化街区渗透。河上增设两座步行桥，与祥符桥距离大于70m，实现对祥符桥视线保护的同时加强了南北岸的交流。为满足社区功能需要，更新方案新建了学校和养老院，并于底部设置与街区联通的开放路径，以推进资源间的共享。

祥符桥传统风貌街区更新以EPC总承包的方式开展，这种设计机构同时作为项目管理方的建设方式对于大型城市更新类项目而言是一种有意义的尝试。

总平面图 / Site Plan

1	祥符桥	1	Xiangfu Bridge
2	茧行	2	Cocoon Warehouse
3	粮仓	3	Granary
4	未来公社及美术馆	4	Future Commune and Art Gallery
5	祥符桥文化街区	5	Cultural & Creative Block
6	养老院	6	Senior Community
7	小学	7	Primary School

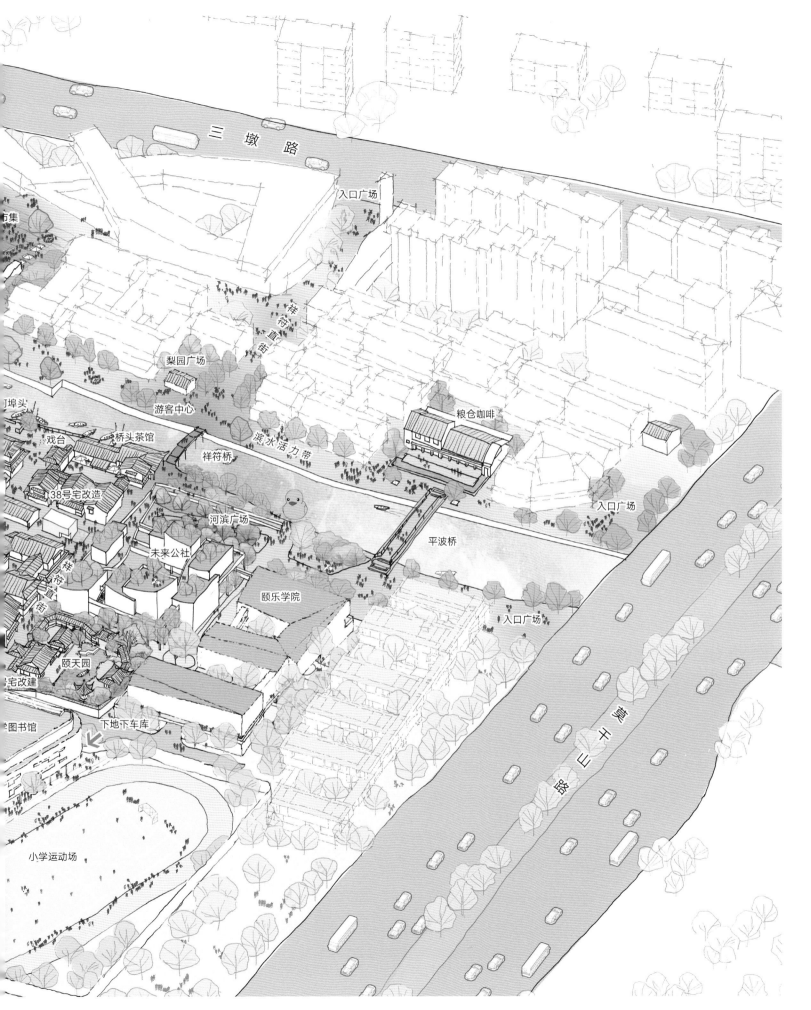

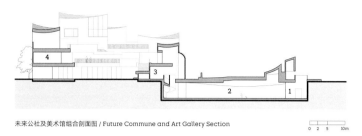

未来公社及美术馆组合剖面图 / Future Commune and Art Gallery Section

1　下沉庭院　　1　Sunken Courtyard
2　美术馆　　　2　Art Gallery
3　美术馆前厅　3　Art Gallery Foyer
4　商业　　　　4　Commercial

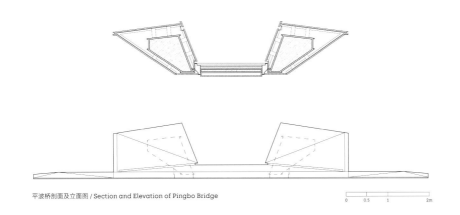

平波桥剖面及立面图 / Section and Elevation of Pingbo Bridge

上海北外滩 32 街坊更新
Shanghai North Bund Neighborhood 32 Renewal

项目地点：上海市虹口区
建筑面积：112,500m²
设计/竣工：2020/—

Location: Hongkou, Shanghai
Floor Area: 112,500m²
Design/Completion: 2020/-

The project's site is a historically preserved neighborhood in Shanghai's Hongkou District. However, the existing buildings within the area had been demolished when the design process began. In response, the design does not merely aim to reconstruct the demolished structures or replicate the previous neighborhood. Instead, it categorizes and scrutinizes the local "pattern language" to uphold the essence of Shikumen-style architecture, offering a revitalized interpretation of the vernacular context.

Strategically, the design original site texture by reviving the original pinwheel-shaped master plan structure, reinstating continuous street interfaces, and adapting the previous architectural composition. It incorporates traditional Shikumen architectural "pattern language" into multiple facets based on the restored fabric, including spatial scale and exterior details. The rhythmic street front is artfully expressed through undulating gable walls, while the iconic Shikumen-style gates and windows enhance the spatial depth and intricacy. Details such as red masonry with gray brick inlays, beige stone molding, iron-crafted railings, and solid brick and stone railings are meticulously retained. For the high-rise residential buildings along the streets, the design abstracts and simplifies the classical tripartite division facade, synthesizing the old and new elements. In the commercial area, an elegant and inviting interface is established by incorporating continuous arches at the ground level.

Beyond exceeding conservation standards, the design fuses classical aesthetics with contemporary materials and techniques in a rejuvenated Shikumen-style neighborhood. This creative approach not only accommodates the demands of modern living but also embraces Shanghai's fashionable identity.

本项目地块为上海虹口区风貌保护街坊，但在设计开端之时，场地内既有建筑已经拆除。面对这一特殊背景，设计没有采用简单复原地块上已拆除建筑或完整复刻某个现存街坊的做法，而是希望通过"建筑模式语言"的归纳和演绎，延续石库门里弄这一建筑类型的内涵与特色，实现"传统石库门语境下的新时代里弄"。

设计辩证性地继承地块的原始肌理特征；维持风车状的整体格局；延续原街区的连续沿街界面；化用原有房屋的拼接组合逻辑。在此基础上，设计于空间尺度、立面形制等各个层面化用传统石库门建筑的"模式语言"：沿街立面以连续山墙面表达韵律感；分层次采用石库门建筑的经典门头形制及开窗形制；沿用红砖嵌灰砖、米色石材线脚、铁艺栏杆、砖石栏杆等细部样式。沿街高层住宅及公寓采用了古典建筑常用的三段式立面，立面整体经语言的简化与抽象传递出新旧交融的气质，连续的拱形塑造出底层商铺的优雅形象。

在满足风貌保护要求的同时，设计从古典审美意趣出发，运用现代建筑材料和技术延续与发展传统建筑形式，塑造了一片适应当代生活需求且具有上海特色的新时代里弄街区。

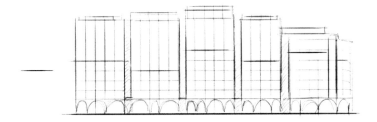

弘安里
Hong'anli

项目地点：上海市虹口区
建筑面积：95,200m²
设计 / 竣工：2021/—
Location: Hongkou, Shanghai
Floor Area: 95,200m²
Design/Completion: 2021/-

Hong'anli is an urban renewal practice located in the 17th neighborhood plot of Hongkou District, an area designated for historical and cultural preservation in Shanghai. The site is intersected by the underground Metro Line 10. The design prioritizes retaining historical authenticity while adapting to contemporary lifestyles, resulting in a vibrant lane-style community that embraces the aesthetic of traditional Shikumen residences.

The site features a collection of lane-style houses representing various types of Shikumen architecture from different historical periods. Three primary and multiple secondary lanes define the overall spatial configuration. Following the existing layout, the design adheres to three guiding principles to ensure the preservation of the vernacular character: ①maintaining the overall configuration of the three primary north-south lanes and multiple smaller east-west lanes; ②retaining the existing fabric predominantly composed of lane-style residences; ③upholding the characteristic of "multi-clusters within one integral neighborhood." These strategies serve as the foundation for the subdivision of land in Hong'anli, allowing new constructions to blend in seamlessly with the historical context.

Before restoration and construction, the existing structures were evaluated and assessed thoroughly. The historically significant Heleli South Two Alleys were preserved in place. Regarding the western street-front gable walls of Heleli North Six Alleys and Baoxingli, and the eastern archway of Heleli North Six Alleys, the roofs were rebuilt, and the remaining plain brick walls were repaired. The degraded moldings on the east facade were restored according to the well-preserved sections. Extensive on-site surveys and research were conducted for other historical structures to ensure that new structures aligned with the original architectural style and maintained the intricate details.

According to on-site environmental vibration testing and structural vibration reduction analysis, a comprehensive steel spring vibration isolator system was installed in all functional rooms to minimize the impact of the underground metro rumble on the upper structures. Steel spring floating floors were also installed in specific parking areas and driveways to dampen structural base vibrations effectively.

总平面图 / Site Plan

1　主弄　　1　Main Alley
2　次弄　　2　Secondary Alley
3　中心花园　3　Central Garden
4　社区配套　4　Community Amenity

本项目基地为虹口区17街坊地块，为上海市第二批历史风貌保护街坊，地铁10号线于地下穿过。设计在改善居住环境的同时，力求实现历史建筑及其环境的延续，基于传统石库门建筑体系诠释适应当代生活方式的里弄式社区。

地块现存多个行列式里弄，由三条主弄与若干支弄构成基本骨架，其沿线留存着不同历史时期、不同类型的石库门建筑。设计在研读地块肌理特征的基础之上，秉持以下三个策略，以延续城市风貌：①沿用三条南北向主弄、多条东西向支弄的总体布局；②维持以行列式里弄为主的肌理类型；③保留"一街坊、多组团"的风貌特征。这三个策略奠定了弘安里的地块划分，新建住宅建筑组团得以充分传承历史的肌理。

在修缮和建造的过程中，场地上的既有建筑被细化甄别和全面评估：对于具有较高价值的和乐里南两弄尽可能地原址保留；对于和乐里北六弄、宝兴里山墙面以及东立面弄口过街门头，拆除屋面搭建，同时修复山墙面清水砖墙，其中后期改建严重的线脚按保存较好的东立面恢复；对于其他的一般历史建筑，基于对建筑现状的勘测以及历史档案的考证，将其原有建筑风格及细节做法应用于新建筑之中。

为减小地铁对上部建筑的影响，在地铁振动环境实测及结构减振分析结果的佐证之下，项目以钢弹簧整体隔振系统完成整体隔振；在部分停车位及行车道区域，设置钢弹簧浮置楼板，实现结构底板减振。

中海顺昌玖里
China Overseas Arbour

项目地点：上海市黄浦区
建筑面积：280,000m²
设计 / 竣工：2020/—
Location: Huangpu, Shanghai
Floor Area: 280,000m²
Design/Completion: 2020/-

The project is situated south of the Xintiandi Business District in Shanghai. The site contains Huangpu District's second batch of historically protected neighborhoods. This high-density community reflects Shanghai's robust commercial and urban characteristics since the city's development as an international trading port. It consists of market houses on the periphery and residential areas with lanes and alleys on the inside. Many existing structures are identified as general historic buildings, while others are listed as protective historical buildings.

The project aims to establish a high-quality urban community with a "New Shanghai style" cultural spirit while preserving the existing architectural style and historical texture. The site is divided into zones for high-rise development and low-rise urban fabric preservation, accommodating residential buildings, commercial areas, and public plazas. Thus, seamlessly integrating the high-rise and low-rise areas for a harmonious coexistence of the new and the old becomes critical.

The master plan maximizes the retention of the "outer market and inner alley" site texture by increasing the preservation area from roughly 47% to around 70%. Building upon the texture, the design introduces public squares and pathways to connect the alley-side commercial areas. A spacious commercial plaza extending inwardly on the north of Shunchang Road integrates public green spaces and serves as the community entrance. Additionally, three historically significant buildings are renovated to house new public functions and are relocated to the site's northwest corner.

The high-rise building facade features aluminum panels and glass, while the lower level uses materials that evoke a sense of age, such as clay masonry, stone, and water-brushed granite. Drawing inspiration from classic Shanghai Shikumen buildings, the architectural form interprets the cultural essence of the "New Shanghai style." Meticulously designed with attention to detail, the railing, floral patterns, rainwater pipes, wind caps, and various other components reflect the proportions and embellishments of traditional architecture.

本项目位于上海黄浦区，北侧紧邻新天地商务区。项目地块内现存街坊为黄浦区第二批风貌保护街坊。街坊内为功能综合的高密度社区，其布局充分体现了上海近代开埠后浓厚的商业城市特征——外部为连续的市房，内部是以里弄住宅为主的居住区。大量现存建筑为一般历史建筑，部分建筑为历史保护建筑。

项目的目标是在最大化保留基地原始的建筑风貌与历史肌理的基础上，实现具有"新海派"文化特色的高品质城市社区。项目用地分为高层区和低层肌理保护区，容纳高层住宅、低层住宅、商业及广场。如何处理高层区与低层区的关系、实现新与旧的共生是设计的重点。

通过总图推敲，设计将原有规划占地约47%的风貌保护范围扩大至约70%，最大化保留了地块"外市内街"的既有肌理风貌。在此基础上，设计适度疏通空间，利用公共广场和巷弄空间将相邻里弄的商业空间串联起来，于顺昌路北侧区域结合公共绿地形成尺度较大的袋形商业广场，同时承担社区入口的作用。此外，设计保留修缮了具有历史文化价值的三处历史保护建筑，平移落位于西北角新址，以全新的业态面向公共开放。

高层建筑外立面由铝板和玻璃构成，低层部分则主要采用陶砖、石材、水刷石等具有年代感的材料。为诠释"新海派"的文化内涵，建筑的形式语言借鉴了上海经典石库门建筑的比例尺度和装饰细节，栏杆、花饰纹样、雨水管、风帽等各类细部构件都有详尽的深化设计方案。

里直街
Lizhi Street

项目地点：浙江省绍兴市
建筑面积：4,300m²
设计 / 竣工：2019/2022
Location: Shaoxing, Zhejiang
Floor Area: 4,300m²
Design/Completion: 2019/2022

Lizhi Street is situated in the old town of Shangyu, where the "Laobadi" weir area holds significant historical importance in canal transportation and preserves the collective memories of the local community. With the development of Shangyu New City in the north, the old town faces challenges such as population decline and regional decay. Therefore, a plan for urban renewal has been initiated to address these issues. Over three years, Lizhi Street has undergone a remarkable transformation, evolving from a traditional market street into a modern block that accommodates retail, exhibitions of historic preservation, academies, and more. Regarded as the "parlor of the old city", Lizhi Street has been revitalized and infused with new vitality.

The approximately 200m-long Lizhi Street becomes a vivid collage of memories. Both sides of the street are adorned with low-rise buildings dating from the late Qing Dynasty to the 1990s. Among these structures are four cultural heritages: Zhenhua Hotel, Dadong Shipping Company, Dawang Temple, and Chen Zonghai Daotai's Mansion.

The plan's guiding principle is the preservation of historical memory. Four cultural heritage relics are preserved unchanged. The old street's telegraph poles, doors, windows, doorplates, streetlights, and slates at building entrances have also been retained and repurposed. In addition to these tangible aspects, the design considers significant site characteristics, such as variations in elevation, the curvature of streets and alleys, and the spatial constraints imposed by courtyard walls. The materials utilized in the project employ modified process techniques. These materials, including dry-stack stone and terrazzo, have historical qualities and are used in the walls, baseboards, window frames, and other architectural elements, evoking the existing material memories.

Based on the existing fabric, The design has strategically guided and enhanced the street interface to provide public spaces that better foster modern commercial activities. For example, after functional modification, the former residential fence with floral-patterned bricks along the street has been repurposed as a setback parapet, providing additional space for the commercial streets. Yet, the original fence's stone base was reused as bench foundations. Similar integration of old and new reverberates throughout the project.

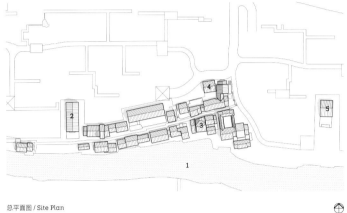

总平面图 / Site Plan

1　萧曹运河　　　　1　Xiaocao Canal
2　原大王庙　　　　2　Former Dawang Temple
3　原大东轮船公司　3　Former Dadong Shipping Company
4　原振华旅馆　　　4　Former Zhenhua Hotel
5　原陈宗海道台府　5　Former Chen Zonghai's Mansion

里直街位于绍兴上虞老城区，其所在的"老坝底"区域与漕运文化有着千丝万缕的关联，承载着老上虞人的集体记忆。随着上虞新城在城市北部的崛起，老城区面临着建筑破败、道路拥塞以及人口外流等问题，城市更新计划就此展开。历时三年光阴，里直街从一条传统的集镇街道转变为容纳现代商业、历史保护建筑展览、书院等功能的特色街道，作为"老城会客厅"焕发出新的活力。

约200m长的里直街承载着一段漫长而混杂的生活记忆，街巷两侧留有自清末民国至20世纪90年代叠加建造的低矮建筑，其中包括振华旅馆、大东轮船公司、大王庙和陈宗海道台府四座文保建筑。

更新方案以留住场所记忆为第一原则。四座文保建筑被原样保留；同时，原街巷中电线杆、门窗、门牌、路灯及入户石板等建筑构件均被保留并加以再利用。在这些实体要素之外，重要的场所特征如场地的高差关系、街巷的曲折角度、院墙与街巷的空间关系等，也被融入更新设计方案之中。设计以改良工艺运用干粘石、水磨石等具有年代特征的建筑材料于建筑的墙面、踢脚线、窗套等部位，以期重现既有的材料记忆。

在旧肌理的底图之上，设计在街道界面梳理出适宜现代商业活动的公共空间。例如，一户民宅的花砖院墙被改造为退后的矮墙，以适应功能置换后商业街开敞空间的需要，但原围墙的石块基础并未被弃用，而是作为公共座椅的承托物，继续见证着街巷的熙来攘往。这类新旧并置的更新手法，在项目各处皆有体现。

For

Living

居住

绿城湖境云庐
Greentown Hangzhou Oriental Villa

仁恒海上源
Yanlord Arcadia

华润亚奥城
CR Land the Century City

绿地海珀外滩
Greenland Hysun Bund

融创长乐雅颂
Sunac Changle Yasong

绿城外滩兰庭
Greentown the Bund Garden

融创滨江杭源御潮府
Sunac Binjiang Imperial Mansion

绿城春风金沙
Greentown Hangzhou Lakeside Mansion

华润武汉瑞府
CR Land Wuhan Park Lane Mansion

蓝城陶然里
Bluetown the Kidult

绿城空中院墅
Greentown Sky Villa

绿城湖境云庐
Greentown Hangzhou Oriental Villa

项目地点：浙江省杭州市
建筑面积：146,100m²
设计 / 竣工：2018/2021
Location: Hangzhou, Zhejiang
Floor Area: 146,100m²
Design/Completion: 2018/2021

As a systematic and fresh approach to low-density duplex residential development, the Oriental Villa exemplifies a synthesis of traditional solidity and modern lightness.

The master plan incorporates the surrounding water bodies into the spatial system, establishing a balance between openness and compactness that extends the natural and cultural splendor of the Xixi Wetland. A central axis connects all architectural clusters. Incorporating the distant river views into the site, the neighborhood center, infinity pool, and expansive green areas create a landscape sequence along this axis. The riverfront is lined with staggered buildings, ensuring all the rear units enjoy lovely water views.

The design of the facade system combines aesthetic appeal with functionality. It integrates space, function, form, and decoration with a unified approach, resulting in a visually capturing expression that encompasses both the interior and exterior of the building. The front facade features extended balconies that create a horizontal visual continuity. These uninterrupted glass interfaces alleviate the feeling of confinement and contribute to the structure's general appearance of lightness. Comprising double-height spaces within the balconies adds a dynamic rhythm to the facade while providing residents with spacious private areas. The wooden balcony ceiling is connected to metal eaves, accentuated by delicate metal bars along the edges. This artistic interpretation of classical composition principles applied to the eave pillars and architrave provides rain protection and imparts a unique aesthetic. Furthermore, the facade system also integrates equipment components. Water pipes are concealed within the spaces between the balconies, resulting in a clean, uncluttered facade appearance.

湖境云庐是将传统与现代融会贯通的一次尝试，也是对于低密度叠拼住宅的又一次系统性革新。现代的轻盈、传统的厚重，在项目中得到统一而充分的表达。

传承西溪湿地的悠然氛围和人文意境，设计借景基地周边水域，创造出疏密有致的地面空间。建筑群落由一条中心轴线串联，邻里中心携无边泳池与开阔绿地形成一串景观序列，将远处的河道景观引入园区内。沿河一线布置相对自由，通过楼栋的错动布局，后排住宅也享有优质的景观视野。

设计探索出一套既满足居住需求、又具备形式张力的立面体系。空间、功能、形式、装饰被整合进完整的秩序之中，建筑从而获得内外融贯的表达。建筑正立面由连续的阳台界面勾勒出强烈的水平线条，连续的阳台界面以玻璃为主材，在自然光线下，透明介质弱化空间边界，消解多层的体量感。阳台局部的两层通高处带来立面节奏的变化，也成为居住者私享的高阔空间。阳台木色吊顶外沿设置金属披檐及纤细的金属杆件，这一做法抽象于古典建筑的檐柱与额枋，遮挡雨水的同时，为建筑赋予独特的审美意蕴。设备构件也被巧妙整合于立面体系中，阳台界面框之间的缝隙恰好用于容纳建筑水管，立面呈现更显纯粹。

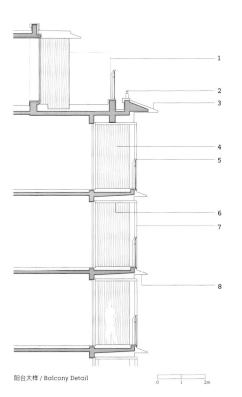

1 金属分户墙
2 铝合金栏杆
3 金属折屋面
4 仿木纹金属格栅
5 六角形铝合金扶手
6 仿木纹金属格栅吊顶
7 六角形铝合金装饰竖杆
8 浅灰色金属吊顶

1 Metal Partition Wall
2 Aluminum Alloy Railing
3 Metal Folded Roof
4 Wood Grain Metal Grille
5 Hexagonal Aluminum Alloy Handrail
6 Wood Grain Metal Grille Ceiling
7 Hexagonal Aluminum Alloy Vertical Rods
8 Light Gray Metal Ceiling

阳台大样 / Balcony Detail

仁源海上源
Yanlord Arcadia

所在地址：上海市杨浦区
建筑面积：292,400m²
设计 / 竣工：2018/—
Location: Yangpu, Shanghai
Floor Area: 292,400m²
Design/Completion: 2018/-

As urbanization continues to evolve, parcel contexts have become increasingly significant in the design of residential communities. It is crucial to consider the interaction between residential structures and the urban environment, as well as their contribution to the overall streetscape. Yanlord Arcadia, as a pilot implementation of the "Living Community" concept developed by GOA, responds to this concern with creative solutions.

The Yanlord Arcadia sits in the heart of the Binjiang residential corridor in the Yangpu District of Shanghai. Jiangpu Park, a community park, is located to the south, while the Yangshupu Port waterfront landscape belt embraces the eastern site boundary. A metro station is only 500 meters away from an industrial renewal parcel across the river. Although the surrounding elements are diverse, they appear dispersed and disconnected.

Drawing upon the context analysis, the architects adopted an urban design perspective for this project. The design breaks from the conventional practice of enclosing residential communities by limiting borders. It transforms the space along the south and east boundaries into inviting open spaces that create seamless connections between the metro station, park, riverfront, and industrial renewal site, thereby establishing a continuous and cohesive urban interface.

Taking cues from the spatial scales and relationships of Shanghai's lilong, the design of the commercial street in the south resembles an engaging "vertical garden." A dynamic and diverse spatial system is produced by overlapping, interspersing, and collaging small architectural volumes. The east boundary incorporates "subtle interventions" such as "transparent" fences and transitional spaces to enhance spatial vitality.

The four residential blocks in the south offer a variety of homecoming routes. Residents can enter via the elevator hall, which connects to the commercial street, or by taking the scenic route around to the second-floor residential lobby, from which they can access the elevator and directly access the exclusive rooftop gardens. This 3D-integrated circulation system immerses residents in urban vitality while maintaining residential privacy.

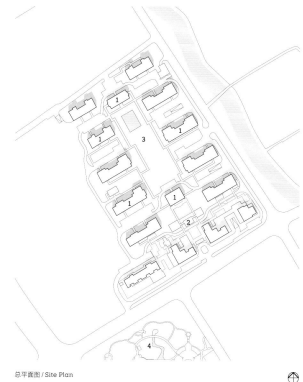

总平面图 / Site Plan

1	高层住宅	1 High-rise Residential
2	花园式商业街	2 Garden-style Commercial Street
3	中心花园	3 Central Garden
4	江浦公园	4 Jiangpu Park

随着城市化进程的深入，错综复杂的地块周边要素成为越来越多住宅地块的设计前提。住宅如何回应城市、贡献于街道，成为一个普遍性命题。作为goa大象设计"鲜活社区"理念的蓝本项目，海上源对此命题予以创造性回应。

本项目位于上海杨浦区滨江生活带的成熟核心区。南侧是杨浦区社区级公园江浦公园，东侧是杨树浦港滨水景观带，隔河相望为杨浦区工业厂区城市更新地块，距地铁站仅500m之遥。周边要素虽完备丰富，却彼此不相连贯，呈现为点状散落的状态。

基于周边环境分析，建筑师选择以城市设计的视角切入。设计开创性地突破了住宅小区以封闭围墙占满边界的常规做法，将南侧及东侧边界的空间塑造为高品质的开放空间，并基于这两条开放边界将地铁站、公园、河滨空间和工业厂区再生地块串联为连续的城市界面。

受上海花园里弄宜人的空间尺度和丰富的空间关系的启发，南侧商业街被塑造为一个"花园式立体街区"，尺度宜人的小体量盒子以叠合、穿插、拼贴的方式形成丰富的空间系统。东侧边界则通过通透式围墙、灰空间等"微处理"提升空间活力。

南侧四座住宅为住户提供多样的归家方式。住户可选择从与商业街衔接的电梯厅直接入户，也可绕行至住区二层的通高大堂，乘电梯直达二层屋顶花园入户。立体化的动线系统帮助每一位居住者在享受城市活力的同时，依然保有居住的私密感和价值感。

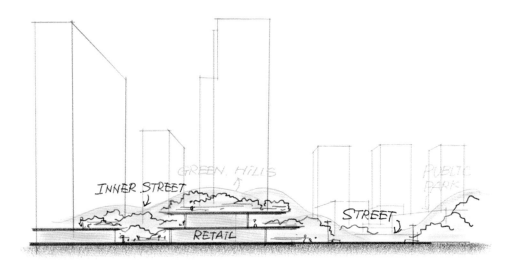

华润亚奥城
CR Land the Century City

项目地点：浙江省杭州市
建筑面积：486,650m²
设计/竣工：2018/2023
Location: Hangzhou, Zhejiang
Floor Area: 486,650m²
Design/Completion: 2018/2023

The project is located adjacent to the Olympic Sports Center and the Qianjiang Century CBD, with its eastern side overlooking the banks of the Qiantang River. During the 19th Hangzhou Asian Games, it will serve as the accommodation for technical officials, complementing the Media Village, Athletes' Village, International Zone and other common areas.

The Technical Officials Village, conceived with the Asian Games as its initial inspiration, transcends its primary purpose to become a multifaceted destination with lasting significance. Beyond the event, the community serves as a future neighborhood paradigm. The design draws inspiration from Hangzhou's natural environment and combines regional and international elements. By reimagining the urban landscape through an abstract geometric architectural language, the village achieves a harmonious synthesis of local identity and global appeal.

The spatial prototype is translated into three levels: architecture, alleys, and clusters. Inspired by the terraced tea fields, the architectural design features a staggered form that offers residents panoramic views of the landscapes. An undulating skyline mirrors the rolling mountains of Hangzhou, infusing the atmosphere with the vibrant energy of traditional alleys. The layout of the apartment buildings, shaped like the letters L and F, enclose well-proportioned courtyards that accommodate residents' desire for sunlight and leisure. Moving between different spaces is like traversing the city's mountains and waterways, each step revealing glimpses of majestic peaks, meandering streams, and serene tea gardens.

The village comprises 12 residential buildings, 10 high-rise apartments, and one super high-rise apartment. Allowing for the creation of diverse and engaging spaces, the apartments are arranged in small-scale blocks. The super high-rise apartment's facade is designed with interlaced pane lines, lending a sense of fluidity and motion to the otherwise static structure and emphasizing the aesthetic appeal of a slender proportion.

本项目毗邻奥体中心和钱江世纪城CBD，东临钱塘江岸。第19届杭州亚运会期间，项目作为技术官员村与媒体村、运动员村、国际区及公共区共同构成盛会的生活配套。

以亚运会为契机，技术官员村被构想为具有国际视野和杭州文化特色的活力中心，并将在赛事后持续为城市周边区域服务。因此，建筑师在社区的空间营造和功能设定上注入诸多创新想法，旨在满足赛事需要的同时，为杭州留下一座标杆级的社区范本。建筑师选择以杭州城的自然环境特征为核心灵感，将城市景观空间原型转译为几何化的形式语汇，以此实现地域性与国际化的相融。

设计对原型空间的转译可具体于建筑—街巷—组团三个层面。错位退台仿佛层层跌落的茶田，使居住者的景观视野实现最大化；在小高层楼栋之间，两侧退台形体形成高低起伏的形态以营造街巷氛围；L形、F形的建筑围合出尺度适宜的院落，满足人们对阳光与活动的需求。行走于社区的不同空间，如同置身于城市山水景观之中，高峰、溪涧、茶园一一展现，移步换景的体验给人以当地自然环境的联想。

整个社区包括12栋住宅、10栋小高层建筑及1栋超高层塔楼。小高层建筑部分以小尺度街区的形式组织于地块中，建筑体量的多样性带来空间营造的潜力。超高层塔楼的竖向铝线条与错层出现的横向线条相配合构成整体立面框架，与"竹节"意象的挺拔俊朗、均衡灵动相呼应。

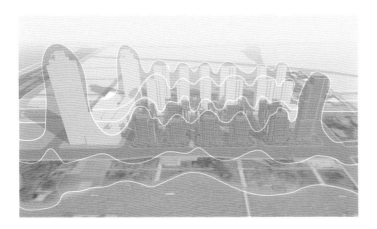

原始体量
Original Volume

双向退台
Two-way Terrace

茗园营造
Tea Garden

景观环绕
Scenery Around

院落全貌
Courtyard Panorama

组团围合
Grouping

花园共享
Shared Garden

院落可达
Accessible

148　居住　For Living

绿地海珀外滩
Greenland Hysun Bund

项目地点：上海市黄浦区
建筑面积：208,000m²
设计 / 竣工：2015/—

Location: Huangpu, Shanghai
Floor Area: 208,000m²
Design/Completion: 2015/-

Hysun Bund stands in the heart of Dongjiadu on the West Bund of Shanghai. The site is adjacent to the 150-year-old St. Francis Xavier Church and the 300-year-old Merchant's Guild, and scattered remnants of old structures, such as traditional courtyards and folk opera stages, surround it. Less than 50 meters from the project site is a 5A-grade super high-rise office building representing the area's summit. Being placed at the crossroads of the old and new poses the design challenge of incorporating the traditional context with the existing urban fabric.

The surrounding environment determines the overall layout. It utilizes an "S" shape to complement the urban texture's rhythm by seamlessly integrating into the urban backdrop, like a well-fitting puzzle piece. This orientation also maximizes daylight access and visibility of the high-rise residences and adjacent office buildings. Residents of the high-rise units enjoy breathtaking landscapes and a close-up view of the north park and historic church from their vantage points. Additionally, looking southeast, they can admire the panoramic vista of the bustling cityscape, featuring prominent landmarks such as the Huangpu River and the Nanpu Bridge.

The project's design features a "wrapping" gesture that creates a sense of unity with the adjacent office building. The sense of volume diminishes due to the primary and secondary massing's layering and intertwining. The use of stone and stainless steel enhances the architectural profile's sleekness. Using a conventional oblique section, the stone elements are sculpted into a distinctive "U" shape to reduce the perceived visual distance and add a touch of elegance to the facade. In addition, titanium-plated stainless-steel accents in gold lines complete the exterior's exquisite details. These accents respond to the Haipai culture of the Bund and contribute to a warm and delicate living environment.

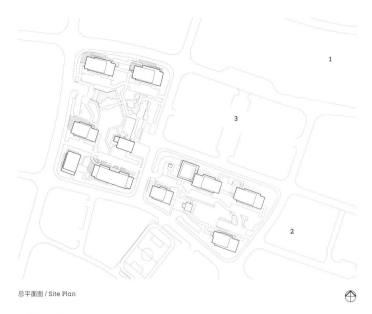

总平面图 / Site Plan

1 董家渡天主教堂 1 St. Francis Xavier Church
2 商船会馆   2 Shangchuan Assembly Hall
3 绿地外滩中心   3 Greenland Bund Center

海珀外滩位于上海董家渡绿地外滩中心内，毗邻距今150多年的董家渡天主教堂和距今300多年的商船会馆，周围零星散落着中式庭院和戏台等老建筑。基地北侧矗立着代表董家渡高度的5A级超高层写字楼，其与基地红线的最近距离不足50m。项目位于新旧交汇点，如何共筑城市界面、容纳新老关系是项目的挑战。

总图秩序由外在环境反推而来，项目如同拼图一般融入城市背景。S形的布局策略顾全了城市区域图底关系的节奏感，既保证了高层住宅与核心办公楼的日照要求，也实现了户型产品视野的最大化。高层住户于中近景处可俯瞰北侧公园和天主教堂，于远景处可遥望黄浦江和东南侧的南浦大桥，将城市繁华印迹尽收眼底。

设计采用与所邻超高层办公楼相呼应的包裹结构：在中心体量之外，穿插渐次跌落的次级体量，化解百米高层的厚重感。边框笔挺的线条由石材与不锈钢材制成。石材由常规斜切的截面形态削减成U形，达到视觉减距的效果，也提升了立面的精致感。由工艺精良的镀钛不锈钢材勾勒出的金色线条是立面的点睛之笔，既回应了以外滩建筑群为代表的海派文化，也传递出温暖柔和的居住氛围。

融创长乐雅颂
Sunac Changle Yasong

项目地点：重庆市巴南区
建筑面积：108,000m²
设计 / 竣工：2019/—
Location: Banan, Chongqing
Floor Area: 108,000m²
Design/Completion: 2019/-

Changle Yasong is situated in the foothills of Qiaoping Mountain, Chongqing. As a reimagining of the Song-style residence and garden, the project bestows timeless values upon a contemporary lifestyle by embracing the aesthetics and deep appreciation of nature inherent in the Song culture.

The Song-garden system emphasizes spatial spirit and natural harmony, featuring a layout with three-sided landscapes and multiple courtyards to enhance natural immersion. By incorporating varying elevations, Changle Yasong's courtyards exhibit a natural balance of openness and seclusion. A private-public sequence establishes multiple extended spatial connections and departs from the inward-focused quality of traditional Jiangnan gardens. This diversified spatial manifestation resonates with contemporary aesthetic preferences. The design depicts three distinct courtyard levels: the delicate "Si Shui Gui Tang" at the entrance, the spacious private yard, and the expansive sky patio. Thus, residents are embraced by verdant beauty daily.

In the annals of ancient Chinese architecture, the Song style is a remarkable fusion of exquisite craftsmanship and elegant simplicity, symbolizing the artistic aspirations and technical achievements of the Song literati. The design reconstitutes traditional architectural elements with a contemporary language. The gilded lines generate the descending posture of bamboo curtains and sunshade levers. The traditional intersecting and contracting balusters are given a modern twist through simplified forms. Furthermore, the second-floor patio incorporates chair-back railings, down-shifting eaves, and abstract decorative brackets, reinterpreting the classical "terrace level." All these adaptations result from a condensed integration of modern techniques and customized designs.

本项目位于重庆樵坪山麓，设计提取和再译宋式民居和园林的特征，将宋代风雅恬淡、尊重自然的风尚重现于今日生活。

与明清江南园林相比，宋式园林更加注重空间的精神性及与自然的关系，常基于三面环景、多重院落的空间格局，呈现出被自然包裹的态势。凭借高差设计，长乐雅颂的庭院展现出天然的开放性及良好的私密性。私人庭院与公共庭院共同构成一组完整的序列，空间之间多有渗透，其意境与明清园林的高墙深院大为不同。设计着重刻画了3个不同层次的庭院——入口处小而精致的四水归堂方院、具有开阔感受的私人庭院及位于居所二层的超大面积观景露台，富于变化的庭院空间将取景自然、疏朗秀丽的园林之美纳入生活日常。

宋式建筑是中国古建筑史上精巧却不繁复的一种建筑形式，这得益于臻于科学的宋代建筑营造技术，以及宋代士族对建筑艺术的理解和追求。运用现代材料和设计语汇，长乐雅颂实现了对宋式建筑和园林的重构与再译，如：采用鎏金色金属线条表达"掰帘杆"和"竹帘挂落"；采用简练的菱形扶手和收分式望柱演绎"寻杖绞角造"；采用阳台美人靠、下移式披檐和抽象装饰斗拱演绎宋式"平座层"等。这些化繁为简的做法背后凝聚了大量专研的现代工艺和定制化设计。

剖立面图 / Sectional Elevation

绿城外滩兰庭
Greentown the Bund Garden

所在地址：上海市黄浦区
建筑面积：100,000m²
设计 / 竣工：2019/—

Location: Huangpu, Shanghai
Floor Area: 100,000m²
Design/Completion: 2019/-

The Bund Garden is located in the Dongjiadu area of Huangpu District, a mere 1.7 km north of the Bund Financial District. Nearby is Laochengxiang (the Old City), a living testament to Shanghai's century-long urban development. The design seeks to create a residential experience that weaves architecture and landscapes while integrating culture and living, drawing inspiration from the historical and cultural setting.

The north facade of the Bund Garden overlooks the intersection of Fuxing Road and the Bund, making it a prominent display surface. Along the site's perimeter, four 100-meter-tall residential buildings maintain a continuous urban interface. The building facade draws inspiration from the elegant architectural forms of Shanghai's Golden Age, infusing it with a contemporary twist. Employing a triadic composition, the facade achieves a dynamic lateral rhythm. The combination of light beige and dark copper aluminum covings adds a touch of modernity while maintaining a classical temperament. The large 2.4m-tall window panels provide residents with breathtaking panoramic views. Guided by a modular system, the architectural details, such as tower covings and corridor eaves, adhere to a consistent design language, celebrating the rigorous aesthetics in classical architecture.

Inspired by the "garden villa" residential prototype in modern Shanghai, the design crafts a garden experience that blends nature

and culture at a near-ground level. Encircling the central garden, several themed small gardens intersect with low-rise buildings, including clubhouses and lobbies. This integration facilitates a smooth transition between indoor and outdoor spaces, providing residents a resort-like ambiance. These courtyards preserve relics of the past, such as ancient trees and remaining architectural elements. The residential lobbies on the ground floor are oriented to face various landscape features, resembling a "floating jewelry box in the landscape."

本项目位于黄浦区董家渡区域，地块北面距外滩金融贸易区仅约1.7 km，西侧毗邻的老城厢是上海一百多年城市发展轨迹的缩影。基于场地的历史文化氛围，设计畅想一种建筑与花园交错、文化与居住融合的生活图景。

项目北侧面向外滩复兴路路口，是城市形象的主要展示面。考虑到城市界面的连续性，4栋100m高的塔楼建筑均沿用地外围布置。建筑立面的灵感来自于上海黄金时代建筑的优雅形式。强调水平线条的三段式构图配合浅米色与暗铜色铝质线脚，赋予塔楼现代与古典兼具的气质；2.4m高的窗扇尺度给居住者以最大化的景观视野。从塔楼线脚到连廊屋檐的建筑细部由一个统一的模数体系控制，以致敬古典建筑的严谨之美。

受上海近代别具特色的人居样本"花园别墅"的启发，设计于近地面塑造自然与文化共融的花园体验。围绕住区的中央花园，若干不同主题的小花园与会所、大堂等低矮建筑相互交错，室内、外空间融为一体，营造出度假酒店般的氛围。场地中的历史记忆——古树、既有建筑构件等元素被保留在不同主题的小花园中。入户大堂面向不同的景观资源，如同"飘浮于风景中的珠宝盒"。

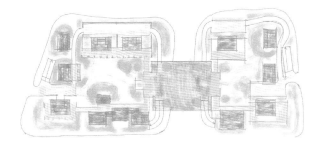

融创滨江杭源御潮府
Sunac Binjiang Imperial Mansion

项目地点：浙江省杭州市
建筑面积：172,700m²
设计/竣工：2018/2022

Location: Hangzhou, Zhejiang
Floor Area: 172,700m²
Design/Completion: 2018/2022

The project's location is where Hangzhou's old city meets Qianjiang New City, just 2.7km from the West Lake scenic area. It is surrounded by a wealth of historical and cultural assets. The master plan achieves a balanced landscape distribution and efficient land utilization, considering the spacing requirements for both slab-style and point-style structures, as well as high-rise and multi-story buildings. All buildings are oriented towards the south, with heights gradually decreasing from north to south and west to east. These height, width, and volume variations add vibrancy to the streetscape and soften the overall project outline.

The elevation of high-rise residences and stacked villas emphasize the sense of geometric composition. The exterior design blends solid and void spaces, accomplished through sleek glass surfaces and balconies framed by aluminum panels in a grid network, while the filleted corners add a touch of warmth to the overall aesthetic. The villas' balconies and terraces offer transparent connections between indoor and outdoor spaces, extending outwards like elegant, streamlined yachts on the ocean.

The design of communal spaces considers various usage scenarios, emphasizing the integration between spaces and activities. The reading area accommodates people of all ages, offering quiet reading nooks and private conference spaces. The sunken courtyard brings ample natural light to the indoor swimming pool and fitness center and maintains good connectivity between the above-ground and underground activity spaces.

本项目位于杭州旧城区与钱江新城的交界地带，距离西湖景区仅2.7km，周边历史人文景观资源荟萃。规划设计将楼栋以正南向布置，根据板式及点式建筑、高层与多层建筑的不同间距要求，自然地组合出错落有致的布局，达成景观资源的均好性和土地利用的高效性。建筑高度由北至南、由西往东逐渐递减，不同楼栋高度、面宽和体量的变化，丰富了地块四周的街道景观层次，也使得项目整体的轮廓线趋于柔和。

高层住宅及叠墅立面皆强调几何元素的构成感。在横向及竖向铝板线条划分出的基础网格之上，平整的玻璃面与阳台构成强烈的虚实对比，细部的圆角处理则传递出温润的感受。为满足室内外空间尽可能通透的需要，叠墅的各标高阳台及露台向外悬挑、舒展，轻盈的流线型令人联想到海潮中的游艇。

社区中的共享空间充分设想了居住者的使用情境，强调空间与空间、活动与活动之间的互融。阅读区既满足全龄段人群阅读的需要，也可容纳具有较强私密性的洽谈活动；室内泳池及健身房通过下沉庭院纳入自然采光，与地上活动空间具有良好的联通性。

叠墅剖透视图 / Stacked Townhouse Sectional Perspective

绿城春风金沙
Greentown Hangzhou Lakeside Mansion

项目地点：浙江省杭州市
建筑面积：239,800m²
设计 / 竣工：2019/2023

Location: Hangzhou, Zhejiang
Floor Area: 239,800m²
Design/Completion: 2019/2023

Lakeside Mansion's design was completed by the end of 2019, an early practice exemplifying GOA's "Living Community" concept. The project is nestled within the Jinsha Lake CBD in the Qiantang New District and enjoys a prime location south of Jinsha Lake Park, separated only by an urban road. At the same time, a short 500m stroll along the lakeside leads to the entrance of Jinsha Lake metro station.

The north parcel interface defines the residential boundary and serves as Jinsha Lake Park's south display, contributing to an accessible urban waterfront area. Instead of erecting a conventional fence, the design introduces a vibrant commercial street. This commercial street contains approximately 6,000m² retail and community facilities. Its spatial arrangement departs from common commercial podiums by incorporating multiple pocket gardens, street corner squares, and second-floor terraces for leisure and relaxation. The main residential entrance is strategically located within the middle section of the commercial street. Following a sequence of "street-courtyard-garden" from the commercial street to the residential entrance, the design enables commercial activities and peaceful living to coexist.

The horizontal layout of the north-facing living rooms maximizes the lake view. Integrating full-height glass windows and balconies connects the lake and the community. The architectural style is reminiscent of the elegant Jiangnan flavor. The light cantilever slabs blur the boundary between indoor and outdoor spaces. Additionally, the commercial podium features pitched roofs interspersed with terraces, fostering an intimate and lively ambiance.

底层商业 / First - floor Commercial

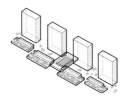

城市客厅 / City Parlor

春风金沙的设计完成于2019年年底，是goa大象设计"鲜活社区"理念的一次早期实践。本项目位于杭州钱塘新区的下沙金沙湖中央商务区内，北侧与金沙湖公园仅隔一条城市道路，沿湖步行约500m即抵达金沙湖地铁站出入口。

地块的北侧边界在定义自身领域感的同时，亦参与塑造金沙湖公园的南边界，是城市滨水开放空间的一部分，具有显著的公共性需求。基于此，建筑师提出"前厅后园"的概念：北侧边界取消围墙，以一条复合型商业街构建住区的"前厅"。商业街内容纳约6,000m²的商业及公共配套设施，其空间格局突破了住宅底商的常规做法，以袋形广场、街角广场和二层露台构建宜人的交往与活动场所，鼓励人的停留。住区主入口位于商业街中段，"街—院—园"的空间序列构建了由动至静、由公共至私密的过渡，商业活动的喧嚣与居住空间的静谧得以和谐共存。

住宅楼栋北面临湖，采用北向宽厅布局，大幅落地玻璃与一步阳台的组合，让湖景与社区无界融合。建筑造型呼应杭州优雅的江南风情，轻盈的水平挑板使得室内空间向外进一步延展。商业的坡屋面与露台穿插，给人以亲切感和活跃感。

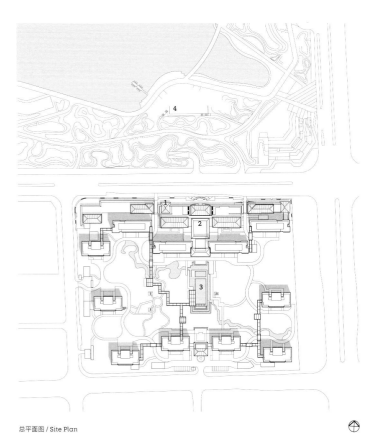

总平面图 / Site Plan

1　沿街商业
2　入口庭院
3　中心泳池
4　金沙湖公园

1　Street-front Retail
2　Entry Courtyard
3　Central Swimming Pool
4　Jinshahu Park

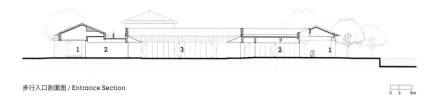

步行入口剖面图 / Entrance Section

1 公共走廊   1 Public Corridor
2 物业办公   2 Property Management Office
3 景观庭院   3 Landscape Courtyard

华润武汉瑞府
CR Land Wuhan Park Lane Mansion

项目地点：湖北省武汉市
建筑面积：298,900m²
设计 / 竣工：2020/—

Location: Wuhan, Hubei
Floor Area: 298,900 m²
Design/Completion: 2020/-

CR Land Wuhan Park Lane Mansion is located in the southern Binjiang International Business District of Wuhan, with the commercial core to the north and the Yangtze River within one kilometer. This exemplary residential development aims to improve the quality of the urban habitat and fabric. Targeted strategies are implemented though the master plan, community form, and defining design attributes.

The master plan highlights a vibrant and multi-dimensional atmosphere that resonates with the dynamic fabric of Wuhan. Arranging buildings of diverse heights in a staggered orientation creates open spaces at regular intervals, resulting in an undulating urban skyline.

Beyond the spatial design, the emphasis is on fostering connections. The design employs a "super plot" strategy to create a comprehensive network of public spaces that integrates the neighborhood with parks, kindergartens, and street-front businesses. This strategy ensures vibrant and engaging social experiences by providing a variety of dynamic and diverse settings.

As a new city landmark, the structures showcase a distinctive architectural language that sets them apart. The towering crown stands as a beacon, symbolizing maritime progress and adventure. The facade's intertwined forms and horizontal lines, reminiscent of the Yangtze River's rhythmic waves, become a metaphor for dynamism and vitality.

华润武汉瑞府位于武汉汉口滨江国际商务区南片，北侧紧邻商务核心区，距长江水岸距离不到1km。作为片区的标杆级居住项目，本项目旨在优化片区人居环境品质，提升城市形态。设计在总体格局、社区营造、标识性三个维度都采取了针对性的策略。

受大开大合的武汉城市格局所启发，本项目整体规划布局强调动态、立体的气势感。风车状总图规划下，不同高度的楼宇交错布置，释放出城市楼宇间通透公共领域的同时，也帮助构建出建筑群落丰富的天际线轮廓。

设计中，比空间更重要的是人的连接。本项目采用"超级底盘"设计社区的公共空间系统，营造沉浸式场景和浓郁的邻里氛围，将社区与公园、幼儿园、沿街商业有机串联起来，为人们提供丰富宜人的交往体验。

本项目作为地标级建筑物群落，采用标识性的形式语言，以跳脱于周遭环境。翩然起翘的超高层塔冠如同城市灯塔点亮高空，带有勇立潮头、扬帆千里的美好期许；利用避难层植入的穿插形体及立面横向线条隐喻长江波涛与水流的气韵。

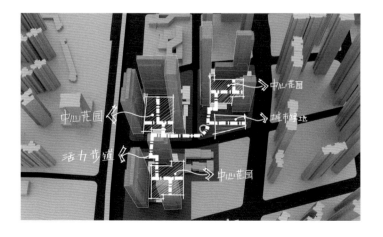

蓝城陶然里
Bluetown the Kidult

项目地点：浙江省杭州市
建筑面积：130,000m²
设计 / 竣工：2018/2023
Location: Hangzhou, Zhejiang
Floor Area: 130,000m²
Design/Completion: 2018/2023

The Kidult is a progressive elderly-oriented community in the Zhijiang Area of Hangzhou. Based on research on domestic and international elderly care models, the project focuses on intergenerational integration, community cohesion, and age-friendly design. Its design goal is to explore a new solution for urban senior care in China's present and future.

A vibrant elderly-oriented community should not solely serve as a residential parcel but should accommodate all generations to stimulate neighborhood vitality. Therefore, the design considers the intergenerational living requirements, from the master plan to the layout of residential units. A shaded avenue separates the neighborhood into two distinct zones. In the northern zone, 3 high-rise apartment buildings are situated along the street, with commercial and community facilities located on the lower levels, forming an age-appropriate vibrant "active zone." On the southern side, 11 low-rise apartment buildings around a central garden create a serene "quiet zone" that offers elderly residents a peaceful and pleasant living environment.

The community promotes abundant social opportunities. The varied landscapes, including a central garden and private courtyards, provide spaces for diverse interactions. Buildings with rooftop gardens provide an ideal setting for elderly residents to enjoy sunny days. According to the property management plan, the senior daycare, community retail, and kindergarten along the street will also be accessible to the surrounding area. These facilities encourage communications within the community and connections between inside and outside environments. As a result, the entire community becomes integrated into the larger urban context.

Age-friendly design is integrated into the architectural details considering the physical changes that occur with aging. For example, the unit layout prioritizes the safety and comfort of senior residents, including considerations for wheelchair maneuverability and accessible bathrooms.

蓝城陶然里位于杭州西湖区之江片区，是一个理念前卫的适老型社区。基于对国内外现行养老模式的研究，设计从混龄社区、邻里共融和适老化设计三个层面展开思考，为中国当下及未来的城市养老模式探索全新的路径。

一个理想的适老型社区不应仅提供老年人的生活场所，更应通过容纳不同年龄段的人群来激发社区的生命力。因此，项目从规划布局到居住单元设计皆考虑了跨代际混居的需要。一条林荫道将园区划分为南北两个区域，北侧沿街的3栋高层公寓与其下方的商业及社区配套共同构成"动区"，丰富的活动形式兼容全龄段人群的需要；南侧围绕中心花园的11栋多层公寓则构成"静区"，低密度的建筑形式为老年人提供安宁、亲切的居住体验。

社区为居住者的交往活动提供充分的可能性。园区景观层次丰富，中心花园与宅间私享庭院的组合满足不同类型交往活动的需要，部分建筑设屋顶花园，为老年人提供晴天晒太阳的理想去处。按照业主的经营计划，沿街布置的护理院、社区商业及幼儿园将对外开放，它们不仅是社区内居民邻里互动的场所，也是社区内、外人群交流的接口，整座社区由此融入于更大范围的城市环境中。

结合老年人不同年龄段身体特征的变化，适老化设计被贯彻于建筑的诸多细节之中。例如，户型设计充分考虑了轮椅回转、无障碍卫生间等适老设计，保障老年人的安全与舒适度。

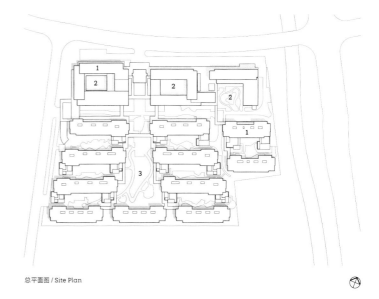

总平面图 / Site Plan

1	娱乐康体中心	1	Recreation and Fitness Center
2	庭院	2	Courtyard
3	中心花园	3	Central Garden

绿城空中院墅
Greentown Sky Villa

项目地点：浙江省湖州市
建筑面积：1,600m²
设计 / 竣工：2020/2022
Location: Huzhou, Zhejiang
Floor Area: 1,600m²
Design/Completion: 2020/2022

The Sky Villa represents a collaborative research initiative between GOA and Greentown China, focused on exploring the incorporation of sky gardens in high-rise residential buildings. The completed prototype is an exemplary model of sustainable design, presenting a new paradigm for coexisting human and natural habitation. The design features a three-dimensional courtyard arrangement, embodying the idea of a "single four-story building with seven courtyards situated in a nine-acre garden." This approach integrates a comprehensive green system into the vertical space. The residential cells strike an ideal balance between interconnectedness and independence by employing split levels and duplex design. Landscape elements extend from the courtyards to the facade, creating an eco-circular system while providing natural sceneries throughout all seasons.

The design implements advanced structural techniques to achieve extending cantilevers. This not only accentuates a lightweight elegance but also creates large sky terraces and transitional areas that blur the boundaries between indoor and outdoor spaces. As a result, residents are presented with numerous opportunities to interact with the surrounding natural environment. Expansive, column-free spaces, open corner glass doors and windows, and 270° uninterrupted glass curtain walls contribute to the seamless integration of interior and exterior spaces and optimize interior luminousness, greatly improving overall living comfort.

In 2022, the Sky Villa attained LEED v4 Gold Certification from the U.S. Green Building Council (USGBC) and WELL Gold Certification from the International WELL Building Institute (IWBI).

空中院墅是goa大象设计与绿城中国合作的一项创新研究课题，旨在探讨高层住宅中空中花园的可能性。已竣工的建筑样板充分展现了可持续性设计理念，为"人与自然共生"的居住理想提供了新范本。设计以"一栋·四层·户七院·九亩园"的立体院落布局，将完善的景观种植系统植入垂直空间。错层结构与复式空间的组合使建筑内的居住空间既相互关联又各自独立；绿色元素从庭院一直延伸至建筑立面，为建筑带来一个四季皆景的生态自循环系统。

在创新型结构技术的加持下，水平向延伸的挑板刻画出建筑的轻盈形象。挑板上下，大面积的空中露台与檐下灰空间模糊了室内外界限，为人与自然的互动创造出多样化的场景。大跨度无柱空间结合开放式转角玻璃门窗和270°无间断玻璃幕墙，既减少了视野阻隔，又保证了建筑的采光性能，有效提升了居住舒适度。

本项目于2022年获得美国绿色建筑委员会颁发的LEED金级认证和国际健康建筑研究院颁发的WELL金级认证。

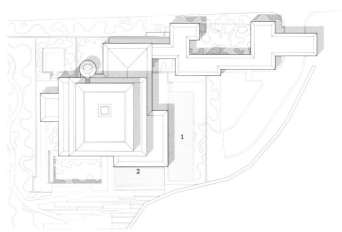

总平面图 / Site Plan

1　泳池　　1　Swimming Pool
2　镜面水景　2　Waterscape

For

Leisure

度假 & 休闲

阿丽拉乌镇
Alila Wuzhen

木守西溪
Muh Shoou Xixi

湘湖逍遥庄园
Xianghu Xiaoyao Manor

杭州远洋凯宾斯基酒店
Kempinski Hotel Hangzhou

苏州狮山悦榕庄
Banyan Tree Suzhou Shishan

德清莫干山洲际酒店
InterContinental Deqing Moganshan

湘湖陈家埠酒店
Xianghu Chenjiabu Hotel

青岛藏马山酒店
Qingdao Cangmashan Hotel

既下山大同
SUNYATA Hotel Datong

阿丽拉乌镇
Alila Wuzhen

项目地点：浙江省嘉兴市
建筑面积：25,000m²
设计/竣工：2014/2018

Location: Jiaxing, Zhejiang
Floor Area: 25,000m²
Design/Completion: 2014/2018

Alila Wuzhen is 3km southwest of the Xizha Scenic Area, proximate to the main road of Wuzhen's new downtown and a picturesque wetland. This location creates an idyllic sanctuary amidst the bustling surroundings.

The resort features 126 villa-style suites. Its master plan adopts the classic Jiangnan settlement as a prototype, drawing from its spatial form, architectural scale, and color relationships. The calm water surface and pure architectural form are harmoniously coordinated, capturing a multilayered and translucent interplay between architecture, architecture and plants, and architecture and water surfaces.

The design reimagines traditional spaces by revitalizing functions and spirits in streets, alleys, courtyards, and other spatial typologies. In addition, the captivating and intricate spatial rhythm of Jiangnan villages has been preserved.

The architectural design embraces an oriental minimalist aesthetic by preserving its essence of simplicity, refinement, and serenity. It enables individual pure geometries to emerge and coexist organically by eliminating redundant decorative elements and employing delicate textures, clean structures, and subtly elegant colors with a modern touch. This conflict between "individuality" and "aggregation" is one of traditional villages' most significant organizational feature.

Alila Wuzhen emerges as a symphony of cultural inheritance and spatial spirit, inviting guests to experience unparalleled charm and immersive wonder. This ethereal architectural cluster gracefully dances between the realms of "imitation" and "reconstruction," exemplifying a fusion of the past and the present. Along the water, the image of a contemporary village appears vividly.

总平面图 / Site Plan

1. 大堂 — 1. Lobby
2. 宴会厅 — 2. Banquet Hall
3. 全日餐厅 — 3. All-day Restaurant
4. 酒吧区 — 4. Bar
5. SPA区 — 5. SPA
6. 健身区 — 6. Fitness
7. 客房区 — 7. Guest Room

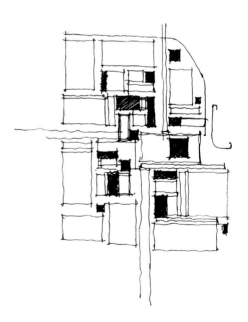

阿丽拉乌镇位于距乌镇西栅景区西南3km处，毗邻新城主干道和一片珍贵的湿地景观，是一座闹中取静的世外桃源。

酒店拥有126套别墅式套房，整体格局以江南村落为原型，提取并再现了传统聚落的空间形态、基本元素、建筑尺度和色彩关系。静谧的水面与纯净的建筑形式相协调，彰显了建筑之间、建筑与植物之间、建筑与水面之间多层次、半透明的关系。

建筑师对传统空间再造的可能性进行了探索。根据度假酒店的使用需求，街、巷、院落等空间类型被赋予了新的功能和场所精神，与此同时，经典江南村落曲折迂回、错落有致的空间韵味依然得以传承。

酒店设计遵循了东方文化中极简的审美倾向。建筑师捕捉和提炼了江南建筑简约、精致、宁静的意趣，摒弃一切直白的装饰性元素，代之以具有现代基调的细腻材料、简练构造以及淡雅色彩，使形式纯粹的单体建筑重复出现，有机生长。这种单一个体和复杂群体的矛盾对立，正是传统村落最为重要的构型特征之一。

在对江南水乡的模仿与再造之间，酒店完成了对地方文化及江南村落空间原型的传承，同时亦契合了酒店对非常规体验和戏剧化场景的特殊需求。在这样一片独特的建筑群落中，历史与当下出现了交叠，一个有机生长的现代村落形象跃然于前。

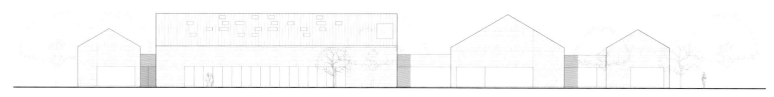

泳池及餐厅组合立面图 / Swimming Pool and Restaurant Elevation

度假&休闲　　For Leisure

木守西溪
Muh Shoou Xixi

项目地点：浙江省杭州市
建筑面积：7,000m²
设计／竣工：2015/2018
Location: Hangzhou, Zhejiang
Floor Area: 7,000m²
Design/Completion: 2015/2018

Muh Shoou Xixi is in the southwest corner of Hangzhou Xixi National Wetland Park, enveloped by native vegetation. It overlooks gentle hills to the south and vast water and plain to the north.

The design converts the five remaining buildings into hotel structures. It adopts a minimal intervention approach that respects the existing vegetation layout by "weaving" the architecture into the landscape, thus celebrating the enchantment of the wetland—its peaceful serenity, rustic wilderness, and secluded sanctuary.

As you enter the site, your gaze is drawn inward to a tranquil water courtyard, with a guqin table and a persimmon tree standing at the center. The water courtyard visually connects the eastern lobby to the western veranda. The veranda, made from reclaimed wooden planks, weathered steel, and water-washed marble, introduces a tangible concept of "time." Every stone and piece of wood within this area bears witness to the ever-changing nature of time through their shifting colors and textures under wind, frost, snow, and rain.

The restaurant is surrounded on three sides by a tranquil waterscape. Its interior combines reused lumber and "stone skin," remnants of lake stone processing. The banquet hall features 270° full-height glazing that merges over 300 m² of indoor space with woods, lawns, and a waterfront terrace. The landscape design subtly infuses landscaping elements that complement the architectural space, following an eco-oriented principle.

The waterway circulation is designed with existing topography. Guests glide through the gentle ripples along the ancient waterway to the hotel, gaining a unique perspective of the wetlands. As they stroll through the wetlands after disembarking, they can spot Muh Shoou Xixi concealed within the forest, revealing the untamed beauty of nature.

总平面图 / Site Plan

1 客房区   1 Guest Room
2 餐厅   2 Restaurant
3 宴会厅   3 Banquet Hall
4 迎宾水院   4 Entry Water Courtyard
5 大堂   5 Lobby

本项目坐落于杭州西溪国家湿地公园西南角。基地南望缓丘山峦，北临水网平原，依水径，通陆路，周围原生植被环绕。

酒店由原有的五栋旧建筑改造而来。建筑师顺应原始植被的布局，以极小规模的介入进行改造，将建筑"编织"到环境中，从而将湿地"冷、寂、孤、野、幽"的自然况味呈现给来访者。

从酒店入口向内望去，一片琴台与一棵柿树静立于水院中央。院落空间以水平线条连接东侧视野开阔的"野堂"与西侧的回廊。回廊以回收的老木板、锈化钢，以及水冲面大理石为主材，借此将"时间"的概念引入空间——当材料的色泽和触感在风霜雪雨中缓慢演变，时间的流逝便也记录在空间的一石一木之中。

餐厅三面临水，以旧木板和湖石加工过程中产生的"石皮"作为室内设计主材。宴会厅以270°落地玻璃围合，逾300m²的室内空间与密林、草坪及亲水平台交融。景观设计同样遵循生态性第一的总体原则，采用轻介入、点式置景的手法，与建筑空间相得益彰。

水路流线的设计基于湿地原始地貌。宾客沿古水道前往酒店，在摇橹船徐徐的桨波中获得观察湿地的另一种视角。下船漫步湿地，可见木守西溪隐匿在密林之中，诠释着这片原始自然的诗意。

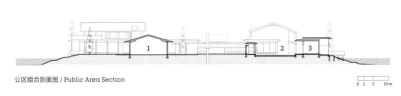

公区组合剖面图 / Public Area Section

1 宴会厅　　1 Banquet Hall
2 迎宾水院　2 Welcoming Water Courtyard
3 大堂　　　3 Lobby

湘湖逍遥庄园
Xianghu Xiaoyao Manor

项目地点：浙江省杭州市
建筑面积：126,700m²
设计／竣工：2015/2019

Location: Hangzhou, Zhejiang
Floor Area: 126,700m²
Design/Completion: 2015/2019

Xianghu Xiaoyao Manor is a large integrated resort with various amenities, including dining options, event venues, recreational activities, youth entertainment, and a total of 318 guest rooms. Located in a serene valley in northwest Xianghu, the resort enjoys a picturesque and natural setting near the lake. However, integrating diverse and dense functions within steep and narrow terrain constraints presents a significant design challenge. Thus, addressing this challenge becomes the starting point of the design process.

The design draws inspiration from mountain villages and adopts a comprehensive planning strategy considering terrain, platforms, and architecture. Embracing the concept of demassification, the architecture resembles a collection of small volumes reminiscent of a mountain settlement. The public areas and guest room clusters are strategically positioned to the east, creating a strong sense of visual tension. Larger functional spaces like the banquet hall integrate into the pedestal, blending with the mountain landscape. On top of the pedestal, smaller spaces with varying heights and gable roofs, such as the hotel lobby and guest rooms, are dispersed, resulting in a dynamic axis relationship that enhances the overall architectural expression. Along the southern foothill, a large expanse of land is preserved as pristine landscapes.

The buildings in different clusters exhibit distinct facade styles. The public areas and guest rooms adopt wooden structural elements found in traditional villages. On the other hand, the small villas incorporate tiles, stones, and rammed earth, showcasing local materials and techniques and enabling a vibrant interpretation of the mountainous wilderness.

本项目选址于湘湖西北角的一处山谷之中。基地形态狭长，山麓陡峭，与主湖咫尺之距，景观视野良好，享有偏安一隅的静谧感。作为大型度假酒店，庄园涵盖完整的住宿、餐饮、宴会、休闲及儿童游乐等五星级服务功能，客房总量达318间。然而"长山谷"与"大酒店"之间存在着矛盾与张力。如何将细致而繁杂的设计需求有机整合于严苛的地形条件之中，成为设计的出发点。

基于对山地村落样本的系统性研究，建筑师对场地进行从地形、台基到建筑的整体规划。在化整为零的设计策略下，酒店呈现为一个由诸多小体量组团构成的山地村落。位于东麓的酒店公区与集中客房是建筑群形态最具张力的部分：宴会厅等大体量功能区藏于下方台基之中，与山体融合；大堂、客房等空间作为一个个高低不同的双坡屋面体量散落其上，呈现出灵活错动的轴线关系。留白的南麓山景则成为纯粹的自然对景。

不同区块的建筑在立面风格上采用了差异化的处理手法。公共区域及集中客房区的建筑着重提取了"传统村落"中的木构元素。小尺度别墅客房则大量采用瓦作、石作、夯土等地域特征鲜明的材料和做法，更加灵活自由地诠释山地野趣。

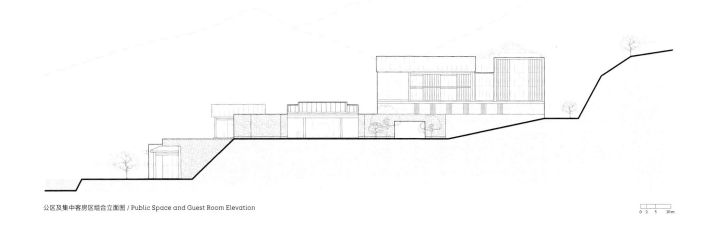

公区及集中客房区组合立面图 / Public Space and Guest Room Elevation

杭州远洋凯宾斯基酒店
Kempinski Hotel Hangzhou

项目地点：浙江省杭州市
建筑面积：60,000m²（地上）
设计／竣工：2012/2019

Location: Hangzhou, Zhejiang
Floor Area: 60,000m² (above ground)
Design/Completion: 2012/2019

Kempinski Hotel Hangzhou, situated adjacent to the premier section of the Beijing-Hangzhou Grand Canal, plays a crucial role in the comprehensive development of Ocean International Center. To the west of the site, Gongchen Bridge, Xiaohe Street, and Dadou Road display rich historical significance, whereas the eastern side is home to the modern Grand Canal Place commercial plaza. The canal's historical memory and the energetic urban environment create a captivating contrast on the site, where the hotel resembles a vessel anchored in a thriving heart, offering an escape from the city's bustle.

The design unfolds a large building volume horizontally, creating a series of staggered landscape platforms to mitigate the overwhelming impact on the riverbank. The lower-level terrace spans 110m in length and 35m in width, housing a full-day restaurant and a swimming pool. Positioned at an elevation of 20m, this terrace offers unobstructed views of the canal surface, extending beyond the lush foliage of street trees. Moreover, the upper-level terraces feature bars that provide panoramic views of the canal as it flows southward.

This architectural terracing creates an undulating skyline for the commercial plaza. A continuous 8m deep canopy defines a well-proportioned vehicle drop-off area along the inner side of the plaza. The main hotel building and the commercial podium are connected in an L-shape, with a corridor extending towards the shopping center in the south.

The architecture features a fully glazed curtain wall system accentuating the facade's transparency. A combination of different sizes and types of glass units creates a sleek and intricate texture, complementing the unique characteristics of each functional block. The guest rooms feature expansive single-pane glass windows measuring up to 6m², the maximum limit for hotel buildings. This design aligns a generous 4.5m-wide landscape view with the room width. Similarly, the all-day restaurant includes quadruple-laminated glass panels, each surpassing 13m² in size.

杭州远洋凯宾斯基酒店毗邻京杭大运河精华区段，是远洋国际中心综合开发项目的一部分。深厚的历史底蕴与鲜活的当代气息交汇于此：向西，是近在咫尺的拱宸桥、小河直街、大兜路等历保建筑与文化街区；向东，则是满载都市活力的远洋乐堤港商业广场。最终的设计如同暂泊繁华闹市的行舟，营造闹中取静的度假体验。

新建酒店如按70m的规划高度建设，将对纵深的河岸景致形成压迫与干扰。因此，建筑师选择将建筑体量按水平向展开，并通过体块的错动形成多个观景平台。低区露台达到110m长，35m宽，全日餐厅、泳池等公共设施沿露台展开，其所处的20m标高位置确保视线能越过茂盛的树冠抵达河面。高区露台设置酒吧，可望见运河南流。

层层跌落的酒店体量同时塑造了商业广场的天际线。面朝商业内广场一侧，进深8m的通长雨棚勾勒出酒店的主入口。裙房商业与酒店主楼呈L形咬合关系，并与南面的远洋乐堤港购物中心通过连廊相接。

建筑主体采用全玻璃幕墙系统以强调立面通透性。不同尺寸和种类的玻璃单元拼合出简洁而细腻的立面肌理，并赋予各功能体块不尽相同的细节特征。客房区6m²的最大单扇玻璃面积达到酒店类建筑的上限，满足了4.5m开间的景观面效果。全日餐厅则采用4层夹胶玻璃，单扇面积超过13m²。

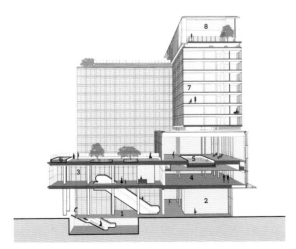

剖透视图 / Sectional Perspective

1	商业街入口	1 Commercial Street Entrance
2	大堂	2 Lobby
3	宴会厅前厅	3 Banquet Antechamber
4	会议室	4 Conference Center
5	泳池	5 Swimming Pool
6	无边景观水池	6 Infinity Landscape Pool
7	客房	7 Guest Room
8	酒吧	8 Bar

苏州狮山悦榕庄
Banyan Tree Suzhou Shishan

项目地点：江苏省苏州市
建筑面积：55,400m²
设计 / 竣工：2020/—
Location: Suzhou, Jiangsu
Floor Area: 55,400m²
Design/Completion: 2020/-

度假&休闲　　For Leisure

The hotel is situated in the southern Lion Mountain Park in Suzhou, looking into Shishan (Lion) Mountain across the lake. It will become integral to the Shishan Scenic Area's landscape system.

Suzhou is celebrated as a garden city for its deeply integrated urban and natural landscapes. Inspired by this urban identity, the design infuses oriental garden aesthetics into a modern hotel. It presents "a garden within a garden" by the lakeside park, weaving internal and external landscapes. The master plan draws a central axis towards the mountain, connecting the entrance, lobby, and all-day restaurant. The guest rooms, featuring villa clusters scatter along the axis's two sides, offer mountain and lake views. The park's water is introduced into the site, further blending the internal and external landscapes.

The Jiangnan settlement creates diverse and dynamic spaces by blending architectural simplicity with experiential richness. Embracing this unique spatial characteristic, the design adopts gable-roof structures ranging from 1 to 3 stories as the primary architectural motif. Roof width, length, and placement variations establish an intriguing interplay between buildings and courtyards. Within a disciplined layout, the architecture encompasses courtyards with valleys, platforms, and waterscapes at different elevations. These courtyards intertwine and nestle together, offering a journey of evolving spatial experiences.

The facades reinterpret traditional aesthetics using contemporary materials. Combining stone curtain walls and hollowed-out brick walls creates an artistic tribute to the classical Su-style fences. The woodgrain-transfer-printed aluminum curtain wall, carved aluminum windows embody a translucent texture, evoking the ethereal atmosphere of a garden bathed in the soft glow of lanterns at night.

酒店位于苏州狮山公园南侧，与狮山隔湖相望。未来，酒店将成为整个狮山公园的一部分，融入其完整的景观体系。

作为园林胜地，苏州以其城园合一而闻名。受此启发，设计意图通过内部造景及对外借景，塑造一个具有传统园林意蕴的当代酒店，一座公园湖畔的"园中之园"。方案以正对狮山方向为主轴线，铺展入口、大堂、全日餐厅等公共功能；集中式及别墅式客房则沿主轴线散落于两翼，面朝狮山与主湖面。外部公园水系被引入场地之内，内外景观浑然一体。

江南聚落之中，形制简单的民居通过组合关系构成丰富多变的空间，建筑的质朴性与体验的丰富性由此得以共存。酒店的设计汲取了其中奥妙，以尺度精巧的1—3层双坡顶建筑为基本单元，通过其宽窄、长短的变化，前后位置的错动构成屋与院交错的图底关系。在秩序感极强的总图布局下，建筑围合出具有不同标高的谷院、台院、水院。这些庭院相互嵌套、连通，给人以步移景异的空间感受。

建筑立面通过现代材料完成对传统意韵的转译。石材幕墙与陶砖镂空墙的组合呼应苏式院墙意象；木纹转印的铝板幕墙、铝花格窗所诠释的半透明的质感，意在致敬夜色之下园林院墅灯影朦胧的意境。

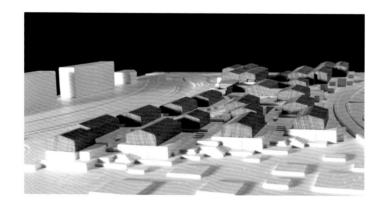

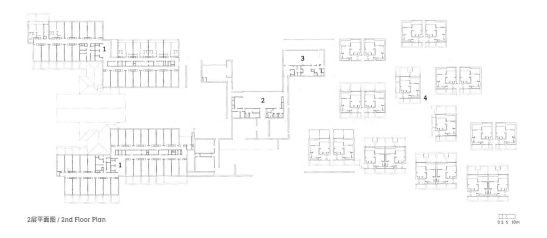

2层平面图 / 2nd Floor Plan

1	集中式客房区	1	Common Guest Rooms
2	酒店大堂	2	Lobby
3	全日餐厅	3	All-day Restaurant
4	别墅式客房区	4	Villa Guest Rooms

德清莫干山洲际酒店
InterContinental Deqing Moganshan

项目地点：浙江省湖州市
建筑面积：53,000m²
设计/竣工：2020/—

Location: Huzhou, Zhejiang
Floor Area: 53,000m²
Design/Completion: 2020/-

度假&休闲　　For Leisure

InterContinental Deqing Moganshan is situated adjacent to the Moganshan Scenic Area. The site is an east-west U-shaped valley, with a meandering drainage canal descending along the mountainside and lush bamboo forests adorning the slopes. Small construction plots scattered across the area, spanning the valley, or occupying steep slopes, present unique challenges for planning and design.

To optimize functional and circulation efficiency, the master plan divides the site into two distinct zones: the upper hillside and the lower valley. The hillside zone accommodates the guest rooms, which require greater privacy and room capacity, while the lower valley zone contains public facilities, including the lobby, all-day restaurant, and swimming pool. The guest rooms are elevated above the ground level to minimize disruption to the natural terrain, while the public areas are integrated into the landscape's contours. The undulating green roofs create an organic blend with the natural backdrop.

The design translates the dramatic site features into unique spatial experiences. One notable example is the executive lounge, a glass bridge that gracefully spans the valley with a hanging full-glazed dining room above water, providing guests with an unparalleled experience.

总平面图 / Site Plan

1 大堂 　　　1 Lobby
2 宴会厅　　2 Banquet Hall
3 餐厅　　　3 Restaurant
4 泳池　　　4 Swimming Pool
5 酒吧 & SPA　5 Bar & Spa
6 客房　　　6 Guest room
7 行政酒廊　7 Executive Lounge

The project makes extensive use of locally sourced materials. Renewable bamboo-based fiber composites lattice on the facade provides effective sun shading and energy-saving solutions while generating natural variations in light and shadow. The bamboo-textured building envelopes establish a connection between the built environment and nature, enhancing the overall sense of tranquility and relaxation that permeates the atmosphere.

莫干山洲际酒店位于紧邻莫干山风景区的山谷内。场地是东西向的U形谷地，一道泄洪渠随山势蜿蜒而下，山坡上竹林遍布。可用于建设的小型地块点状散落其间，或横跨溪谷，或地处陡坡，给酒店的规划设计带来挑战。

从建筑功能和流线角度出发，用地被划分为谷地区与坡地区。对私密性及功能容量要求较高的酒店客房被设置在高处的坡地区，而酒店大堂、全日餐厅、泳池等开放性较强的功能则被设置于谷地区。设计采取了一系列体量消隐策略：坡地区的客房以架空形式减小对地表的干预；谷地的公共区域则结合所衔接道路的标高，以地景式做法延展于地面；覆绿屋面在自然背景中曲曲折折，与之融为一体。

设计将戏剧化的场地特征转化为令人难忘的场景体验。例如，行政酒廊被设计为一座跨越谷地水系的玻璃桥，一只可升降的玻璃盒子作为特别包厢悬放于溪谷之上，为宾客提供极致的用餐体验。

项目大量使用本地建材。建筑外立面采用可再生的竹钢作为主材，其制成的格栅构件不仅起到遮阳节能的作用，同时也带来曼妙的光影变化。覆以竹钢格栅肌理的建筑物与周围的竹林环境相得益彰，增添了建筑整体的度假氛围。

1 景观水池	1 Water Feature
2 行政酒廊	2 Executive Lounge
3 观景厅	3 Landscape Cube
4 特色餐厅	4 Restaurant
5 厨房后勤	5 BOH
6 观景平台	6 Landscape Terrace

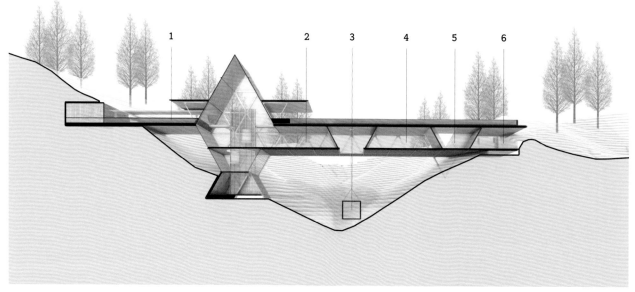

行政酒廊剖透视图 / Executive Lounge Sectional Perspective

湘湖陈家埠酒店
Xianghu Chenjiabu Hotel

项目地点：浙江省杭州市
建筑面积：58,000m²
设计 / 竣工：2015/2023

Location: Hangzhou, Zhejiang
Floor Area: 58,000m²
Design/Completion: 2015/2023

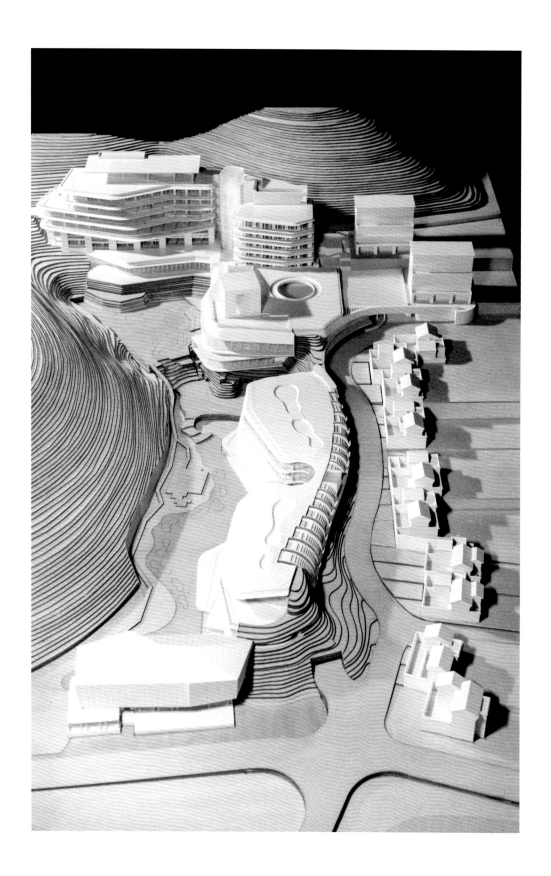

度假&休闲　　For Leisure

The project is located in the Xianghu Tourist Resort Area in Hangzhou. The site is long and elongated, stretching deep into the valley of Jiangjun Mountain, and is surrounded by mountains on three sides. An approximately 30m elevation difference exists between the site's eastern and western portions.

The design conserves the Xianghu Tourist Resort Area's mountainous topography by adopting a "subtle architecture, strong landscape" approach. This approach utilizes the site's natural features, including sloping terrain, cascading waterfalls, rugged cliffs, and abundant vegetation, to create a cohesive valley landscape system. Within this system, public facilities such as restaurants, event spaces, a swimming pool, and a fitness center are strategically distributed, harmoniously blending with the surrounding environment.

The hotel lobby is situated at an elevation of 21 meters in the middle section of the valley, offering breathtaking views of the surrounding landscape. The 30m-tall atrium serves as the vibrant centerpiece of the hotel and separates the guest rooms into two clusters: the lower rooms provide picturesque vistas of the southern Jiangjun Mountain, while the upper rooms offer panoramic views of the serene lake and distant mountains. An extensive circulation system connects these two types of rooms to form an integral organism.

The metal and glass facade mirrors the surrounding environment, while the main structure is harmoniously embedded within the mountain, seamlessly merging with the natural landscape and its visual impact. The interior design aims to create a consistent and immersive spatial experience for visitors. From the lobby to the courtyard of the lower-level guest rooms, the double-height interior spaces are filled with an interplay of natural light and scenic views. Ramps connecting different levels and a glass elevator form a walkable space that is both enjoyable and functional.

本项目位于杭州湘湖旅游度假区。基地形状狭长，深入将军山谷地，三面群山环抱，东西高差近30m。

建筑师希望在不破坏原有山形地貌的基础上，以"弱建筑、强景观"的态度回应湘湖景区的山水肌理。因此，设计利用山坳地形，以错落的叠瀑、嶙峋的岩壁及丰富的植被构建起连贯的峡谷景观系统，并将酒店的餐饮、聚会、泳池、健身等公共功能散布其间。

酒店大堂位于峡谷中段21m标高处，拥有别具特色的景观视野。近30m通高的中庭作为酒店的交通枢纽，将客房区分为山下和山上两部分——山下客房以南侧的将军山为主要对景面，山上客房则以远处湘湖及对岸群山为主视野。两种类型的客房通过复合式交通纵横相连，成为有机整体。

建筑的开敞面运用金属和玻璃映射周边环境，主体则嵌入山体之中以弱化体量感，模糊建筑与自然的界限。建筑的内部空间意在构建连续的感知体验。从大堂到山下区客房中庭，地景和天光在通高空间交汇，坡道贯穿各层，垂直景观电梯于山谷间穿梭，一系列场景共同编织出独具魅力的空间印象。

青岛藏马山酒店
Qingdao Cangmashan Hotel

项目地点：山东省青岛市
建筑面积：20,260m²
设计/竣工：2018/—

Location: Qingdao, Shandong
Floor Area: 20,260m²
Design/Completion: 2018/-

The project is situated at the sloping foot of the majestic Cangma Mountain, with a total elevation difference of about 30 meters, and overlooks the tranquil water at the southeast. Its design presents a "Mountain Village on Terraces," delivering a heavenly retreat.

Following the philosophy of "hidden elegance in simplicity," the hotel reveals itself as terraced fields that merge with the surrounding mountains, with each floor following the contours of the mountain slopes. Its public area adopts an L-shaped layout, positioned to encircle the northern mountain in a staggered manner to achieve borrowed scenery. Guests can enjoy stunning views of either the majestic mountains to the north or the serene lake panoramas to the south. By integrating cultural and artistic themes into their design, these public spaces provide a deeper sense of spiritual fulfillment.

The banquet hall of Cangma Mountain Hotel opens to nature like a forest cottage, unlike those found in cities. Its upper section features lucid glazing, while the lower part incorporates openable doors and windows. The restaurant and book lounge on the upper level provide a scenic view, while the southern outdoor terrace adds a picturesque foreground to the lake vista. The stacking of 52 guest rooms provides different nature perspectives at various heights. Each has two spacious private courtyards, and every window enjoys a frameless wild scenery.

The design encapsulates the essence of Qingdao's rich history and Cangma Mountain's breathtaking scenery by depicting a stone settlement imagery. The facade crafted from Shandong's indigenous granite and the concise gable roofs further enhance local identity, symbolizing the unpretentious charm of a traditional mountain village.

酒店坐落于壮丽秀美的藏马山南麓，位于整体高差约30m的坡地之上，东南侧面向静谧的水面。设计以"山居梯田村落"为题塑造一片世外桃源。

在"藏巧于拙"的设计理念下，酒店依山就势，各层建筑都与周边山体顺畅衔接，以最轻柔的方式贴合山地，隐现于自然山水之间。酒店公共区域呈L形布局，围绕场地北侧山体，通过错动实现了空间各部位的借景：或北看山体、或南眺湖面。公共区域在空间运营上纳入文化艺术概念，山水景色之中给人以更为深度的精神享受。

有别于传统意义上封闭的都市酒店宴会厅，藏马山酒店宴会厅面向自然开放，如同一座森林中的木屋；高处为通透的玻璃，底部为可开启的门窗。餐厅及书吧居高临下，南侧的室外平台为此处的赏湖视线添加了动人的前景。52间客房层层跌落，提供不同标高之下各具况味的观湖体验。每一间客房均拥有前后两处宽敞的私密院落，每一扇窗都享有超宽视野的无框山野画卷。

设计将青岛的百年历史文脉和藏马山壮丽的石山风景凝练为石山村落的意象。建筑立面以山东原生花岗岩为主要材料，配以简洁的双坡屋顶形体，诠释了山地村落的质朴与低调。

总平面图 / Site Plan

1	宴会厅	1	Banquet Hall
2	大堂	2	Lobby
3	餐厅	3	Restaurant
4	酒吧及书房	4	Bar & Book Bar
5	客房区	5	Guest Room
6	贰号酒店	6	No.2 Hotel

既下山大同
SUNYATA Hotel Datong

项目地点：山西省大同市
建筑面积：9,800m²
设计/竣工：2021/—
Location: Datong, Shanxi
Floor Area: 9,800m²
Design/Completion: 2021/-

SUNYATA Hotel Datong is a 45-room boutique hotel adjacent to Heyang Gate. The site comprises historically protected structures on the west, occupying approximately 1/3 of the entire site. As remnants of the ancient city blocks, these structures contribute to the hotel's distinctive character and imbue it with a profound historical charm.

The design is inherent in the spatial prototype of the Shanxi courtyard while weaving old and new elements into Datong's historic fabric. It embraces the traditional Shanxi residences, known for their grand exteriors, intricate layouts, and sophisticated aesthetics. The new section follows classical courtyards' scale, architectural composition, and color schemes. The local living heritage is honored by preserving and replicating local features like the "heshe" (courtyard house), "yongdao" (corridor), and "guojielou" (bridge building). In the central courtyard, one can look towards the west and see the old residences renovated into guest rooms. They blend in with the modern structures, creating a dialogue between the past and present.

The hotel entrance features a sunken courtyard that leads to the Yungang Art Museum, which includes exhibition halls, a theater, a mural restoration workshop, and a café. The underground museum and swimming pool are illuminated by a skylight, evoking a serene and rustic ambiance reminiscent of natural caves.

既下山大同位于历史悠久的大同古城东侧，毗邻和阳门，周边有代王府、法华寺、善化寺、九龙壁等历史建筑，是拥有45间客房的精品酒店。位于西侧的历史保护建筑占据场地约1/3面积，它们作为古城里坊的残片为项目带来厚重的历史特质。

设计基于对山西院落空间原型的传承，通过新旧织补的方式延续大同古城的里坊肌理。山西大院民居的精髓被凝练为"外雄内秀、深藏不露、别院深深、五进穿堂"的抽象特征，渗透在新建筑的蓝图之中。新建部分在院落尺度、建筑组合、色彩关系上借鉴了传统民居意匠，合舍、甬道、过街楼等独属于当地的场景被保留和再现，向旅人传递大同的传统生活记忆。隔着中心的庭院向西望去，原貌复原后作为客房区使用的古宅与身畔的新建筑构成了一场今昔的对话。

酒店入口处特设的云冈艺术馆包含数字化展厅、小剧场、壁画修复工坊、下午茶空间等，是颇具本地特色的惊喜之处。设计从融合于大地的石窟景观得到灵感，以一个下沉式院落组织从地上延伸至地下的艺术馆空间。在顶部天光的渲染之下，地下艺术馆及泳池给人以天然洞窟般的质朴体验。

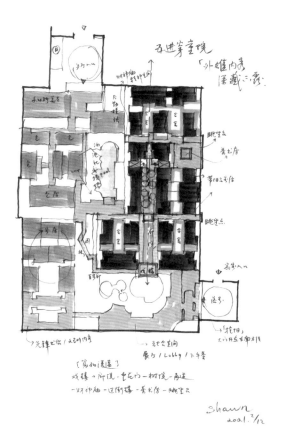
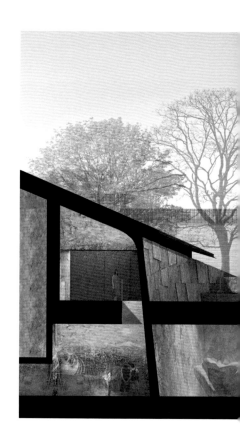

For

Suburban-ur
Mutualism

城乡协同

阳羡溪山
Yangxian Landscape

曲水善湾乡村振兴示范区
Qushui Shanwan Rural Revitalization Demonstration Area

曹山未来城古桥水镇
Caoshan Future City Guqiao Water Town

张謇故里小镇柳西半街
Jianli Town Liuxiban Street

阳羡溪山
Yangxian Landscape

项目地点：江苏省无锡市
建筑面积：635,000m²
设计 / 竣工：2017/—
Location: Wuxi, Jiangsu
Floor Area: 635,000m²
Design/Completion: 2017/-

Yangxian Landscape, a comprehensive town development project, is situated 10km from Yixing City in a rural area. The site's picturesque landscape blends flat plains and undulating hills. The design generates a master plan for nearly 4 km² of land, featuring a harmonious mix of programs, such as courtyard residences, multi-story senior apartments, cultural venues, commercial spaces, and service facilities. The expansive landscape has been carefully preserved, developed, and transformed into sustainable farms, nature reserves, and eco-parks.

Besides fully preserving the existing ecological system, the design creates a comprehensive layout that follows a "three vertical and three horizontal" concept, harmonizing with the natural terrain. Using this spatial framework as a foundation, the architects adopt the "Land Mosaics" philosophy, weaving architectural clusters into the landscape for a harmonious coexistence between human habitats and nature.

The master plan features a fence-free slow-mobility system in which the landform becomes the natural boundaries. In addition to essential amenities such as housing, healthcare, and commercial facilities, the development features an extensive selection of recreational amenities for all ages. These include a mine-pit park, a wellness park, a valley playland, libraries, academies, and cycling courses, ensuring that residents and visitors can enjoy delightful experiences throughout the town anytime.

Several small-scale venues stand out for their unique qualities. In the design of the mine-pit park, for instance, the architects preserved the mine pit as a memory of industrial civilization while transforming the surrounding tea fields into a natural boundary. Within the park, a sports center nestles against the cliff, with a swimming pool that extends from the second floor. The tranquil water surface mirrors the nearby rocks and the distant mountains and sky. Immersed in this serene ambiance, visitors are invited to explore leisurely, meandering between the cliffs and the mountains.

总平面图 / Site Plan

1	雅达书院	1	Yada Academy
2	小镇中心	2	Town Center
3	雅达剧院	3	Yada Theater
4	矿坑体育公园	4	Mine Sports Park
5	雅达医院	5	Wuxi Yada Hospital
6	松下社区	6	Panasonic Community
7	小镇副中心	7	Town Sub-center
8	低密社区	8	Low-density Residential Community

阳羡溪山是一个综合型的小镇开发项目。基地位于距离宜兴市仅10km的乡村中，具有平原与丘陵交织的地貌特色。建筑师对近4km²的用地进行了整体规划。整片区域是包括院落式住宅、多层养老公寓、公共文化设施、商业服务设施在内的综合型小镇；同时，大面积的既有景观资源得到了科学的保护与开发，成为星罗棋布的农庄、公园、乐园等。区域不仅满足常住者的生活需要，更吸引游客前来旅居。

设计团队对每一片山林进行坐标定位，最大化保留原有生态如林地、茶田等，依山就势构建"三纵三横"的整体格局，在这个空间体系的基础上，建筑师遵循"土地嵌合体"理论，将建筑组团镶嵌于自然中，使得人居场所与生态环境密切交织。

小镇置入全域开放的慢行道路系统，所有区域均不设围墙，以地貌地形作为天然的边界。矿坑主题公园、健体公园、骑行游线、山谷乐园、书院等全龄文化娱乐设施分布其中，协同生活服务配套、医疗配套和商业配套，为旅居者提供全域、全时的体验。

项目中的小型公共建筑特色十分鲜明。例如，在设计矿坑公园时，建筑师将矿坑作为工业文明的见证和印记保留了下来，以矿坑周边的茶田作为功能区域的天然边界展开新的建造。其中的体育中心背靠矿坑崖壁，二层的游泳池从室内一直延伸至室外，置身其中，平静的水面倒映出近处的岩石、远处的山峦与天空，人如在崖壁远山间游弋。

小镇中心总平面图 / Town Center Site Plan

1 中心广场　　1 Central Square
2 商业街　　　2 Commercial Street
3 紫砂会馆　　3 Purple Clay Guild
4 艺术家工作室 4 Artist Studio

曲水善湾乡村振兴示范区
Qushui Shanwan Rural Revitalization Demonstration Area

项目地点：江苏省苏州市
建筑面积：1,300m²
设计/竣工：2020/2022
Location: Suzhou, Jiangsu
Floor Area: 1,300m²
Design/Completion: 2020/2022

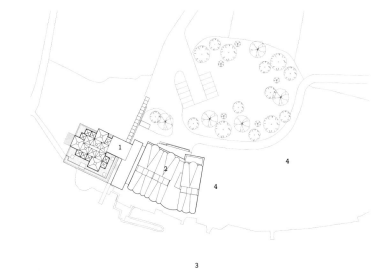

1	水杉居餐厅
2	船坞民宿
3	钟家荡
4	保留村宅

1	Restaurant of Metasequoia Grove
2	Boatyard Hotel
3	Zhongjiadang Lake
4	Existing Housing

总平面图 / Site Plan

Shanwan Village is situated in the Zhongjiadang region, a prominent pilot area for Demonstration Zone of Green and Integrated Ecological Development of the Yangtze River Delta. Since 2020, the region has implemented rural revitalization initiatives to promote the agricultural, cultural, and tourism sectors. Nestled along the expansive Taihu Lake, this small village depicts a romantic epitome of the Jiangnan landscape. The Restaurant of Metasequoia Grove and the Boatyard Hotel are small structures at the village entrance, catering to enjoyable vacation experiences.

A small metasequoia grove by the site inspires the restaurant's design. They are translated into a minimalist architectural expression, a frustum geometry. Each frustum edge is proportioned in a 2:3:4 ratio to create a continuous canopy resembling an artificial forest silhouette. The skylight atop the roof unit allows soft natural light to permeate the interior spaces. The canopy system comprises three layers: the outer layer features custom perforated aluminum panels resembling a branch texture, the middle layer consists of glass, and the inner layer showcases a wooden lattice pattern.

The inspiration for the hotel comes from a natural wharf to the south of the site. The design translates the awning boat's arched canopy into an undulating roof to compose a folk rhyme. Beyond adding a new dimension to the gentle rural skyline, the image of clustered awnings evokes the fancy for thriving beauty. As boats gather, stories of reunion unfold. The hotel's dining area sinks 0.3m below the water as if people were sitting in a boat cabin containing the memories of water town residents.

Within the idyllic setting of water, sky, and the village intertwining in a tranquil panorama, the new restaurant and hotel emerge as a romantic touch to the landscape.

苏州吴江善湾村所处的钟家荡周边作为"长三角生态绿色一体化发展示范区"的先行启动区，自2020年来已开启一系列以农文旅产业为主导的乡村振兴计划。作为太湖沼泽平原之上典型的江南村落，善湾村傍水而建，景致优美，水、天、村落构成了一幅舒展的背景画卷。水杉居餐厅及船坞民宿组成的村落入口组团，在承接度假人群需求的同时为村庄提供崭新的活动设施。

水杉居餐厅的设计由基地周边的一小片水杉林而起。水杉的形象被转译为抽象的几何形态——方锥体，这些方椎体单元基于底面2:3:4的边长关系组合为一整片的连续屋顶，如同自然界中的"众木成林"。屋顶单元顶部的采光窗引入温和的自然光线。屋面系统分为三层：外层为定制穿孔铝板，呈现水杉的枝干纹理；中层为玻璃；内层则使用木格栅。

船坞民宿的灵感来自场地南侧的天然水埠码头。设计提取乌篷船拱形顶棚的形体特征，转译为建筑的拱形屋顶。在船只汇聚的时节，源于聚集的故事就会发生。船篷拥簇的意象表达，激发游客对于水乡繁荣胜景的遐想，也为天际线相对平坦的乡村带来新的建筑几何元素。民宿的餐厅处于一层的临水面，占据着绝佳的观景位置。餐厅的地面相较水面有0.3m的下沉，游客置身其中，仿若进入承载水乡居民旧时记忆的乌篷船舱。

餐厅与民宿在为人提供观景场所的同时，本身也成为点缀于岸线的浪漫景致。

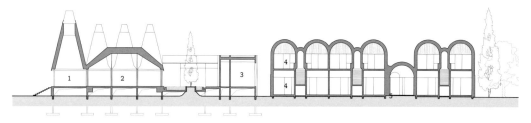

餐厅及民宿组合剖面图 / Restaurant and Hotel Section

1	檐下灰空间	1	Under-eave Transitional Space
2	餐厅 & 村民议事厅	2	Restaurant & Assembly Hall
3	茶室	3	Tearoom
4	客房	4	Guest Room
5	走廊	5	Corridor

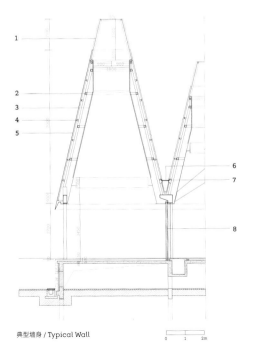

1	中空玻璃	1	Insulating Glass
2	钢结构龙骨	2	Steel Keel
3	内装饰格栅	3	Interior Decorative Grille
4	隔热金属型材窗框	4	Heat-insulating Alloy Window Frames
5	穿孔铝板	5	Perforated Aluminum Sheet
6	可踩踏排水箅子	6	Drainage Board
7	金属搭接盖板	7	Alloy Lap-jointed Ceiling Board
8	中空玻璃幕墙推拉门	8	Insulating Glass Sliding Door

典型墙身 / Typical Wall

曹山未来城古桥水镇
Caoshan Future City Guqiao Water Town

项目地点：江苏省常州市
建筑面积：171,000m²
设计/竣工：2019/—

Location: Liyang, Jiangsu
Floor Area: 171,000m²
Design/Completion: 2019/-

Caoshan Future City sits adjacent to the Caoshan Scenic Area in Shangxing Town, Liyang City. It is designed to integrate various commercial, leisure, artistic, and recreational programs to foster local economic growth through cultural tourism and exhibition opportunities. As the project's heart, Guqiao Watertown offers a distinctive setting with its hospitality and commercial clusters, providing visitors with a unique and memorable stay.

The site has a variation in terrain elevation and a dense network of natural waterways. Five clusters revolve around the concentrated landscapes of the central lake area and the wetland island groups. These include Hillside Accommodation, a Song-style Waterfront Street, Island Residence, River Alley Accommodation, and Central Commercial District. This arrangement creates a low-density, ecological, and organic settlement.

The design process began by thoroughly analyzing the site's topography and geomorphology. To ensure an uninterrupted river view, the site's benchmark water level was carefully aligned with the normal water level of the Dabafang Reservoir in the south. The water routes within the River Alley Accommodation cluster were intentionally shaped into elongated forms. Adopting the existing variation in terrain elevation, the Island Residence cluster contains diverse island landforms, including wetlands, terraces, and tranquil water.

Considering the user demands, the design proposes creative spatial concepts for vacation residency clusters. The boutique hotels are designed as compound units in a combination of "front retail spaces + rear guest rooms" or "ground-level retail spaces + upper-level guest rooms." Their street interfaces feature canopies that invite people to linger beneath, while courtyards can serve as outdoor commercial seating areas, accommodating future operational desires. Consequently, the accommodation clusters possess remarkable potential for cultivating a sense of community through operations. Local small hotels can collaborate and share public spaces like bars, bookstores, and coffee shops. This strategy embraces the intimate nature of boutique hotel architecture while maintaining operational efficiency and a personalized vacation experience.

曹山未来城位于溧阳上兴镇，毗邻曹山风景区，是一个集会议、娱乐、创新、休闲度假为一体的产业新城，旨在以文旅与会展为契机带动当地经济的发展。古桥水镇作为其中的核心区域，将容纳富有特色的民宿及商业聚落，为人们提供与众不同的旅居体验。

场地具有一定高差，密布天然水网。整体规划围绕中心湖区及湿地岛群的核心景观，铺陈坡地民宿、宋韵水街、岛居民宿、水巷民宿、核心商业五个组团，形成一片生态有机的低密度聚落。

设计的开端是对场地地形地貌的再梳理。场地内水面的基准水位被设定为与南侧大坝坊水库的常水位平齐，以实现水域景观的开阔性。水巷民宿组团内，水系被规划为富有纵深感的窄长状；利用原始地形高差，岛居民宿组团内构造有湿地岛、坡地岛、静水岛等多种岛屿地貌。

从目标人群需求出发，设计对度假旅居的新型空间载体"民宿集群"展开创新性设想。民宿建筑被设计为前店后宅或下店上宅的复合型单元，底层沿街处设置可供人停留的檐廊、可拓展商业外摆的小院，满足后期商铺的经营需要。由此，民宿建筑群得以具备"公区共享"的运营可能 —— 一个区域的民宿品牌可以联合起来，共享各自的酒吧、书吧、咖啡店等公共区域。这种方式适应了民宿建筑小微化的特征，能够在保证运营水平的同时兼顾体验的个性化。

总平面图 / Site Plan

1 大坝坊水库	1 Dabafang Reservoir
2 中心湖区	2 Central Lake Area
3 湿地岛群	3 Wetland Island Group

张謇故里小镇柳西半街
Jianli Town Liuxiban Street

项目地点：江苏省南通市
建筑面积：14,600m²
设计 / 竣工：2019/2021
Location: Nantong, Jiangsu
Floor Area: 14,600m²
Design/Completion: 2019/2021

The project is situated in Nantong, the hometown of Zhang Jian, the modern Chinese industrialist. Spanning about 4.7km², this rural revitalization practice focuses on renewing an ancient town via an "industry + cultural tourism" pattern. At the heart of the town, Liuxiban Street is developed in alignment with the overall renovation plan of Changle Old Street, which dates to the Qing Dynasty. The design involves the restoration of six cultural heritages and four historically protected structures and preserves old street memories like flagstone pavements. Commercial and cultural arts programs intertwine with the historical elements to create a delightful mix of streets, plazas, and courtyards on pleasant scales, offering a cultural journey that resonates with the vernacular.

A new exhibition hall, situated at the eastern end of the old street, stands as the cultural landmark of the entire town. Adjacent to the exhibition hall, on the north side, lies the Changle Town Supply and Marketing Cooperative, a century-old wooden building under preservation. The exhibition hall is inspired by the local residential prototype of the "moated dwelling." It divides one roof piece into four elegant single-sloped double-curved surfaces. This decentralization harmonizes the new large-scale building with the old low-rise structures, resolving the scale conflict. The converging roof pieces define a narrow courtyard, enlivening the exhibition spaces with natural light and shadow variations.

The exhibition hall employs a steel structural system with columns integrated in the walls, resulting in a spacious column-free interior. Each roof structure has an "air cushion" cavity with two slender ends to achieve visual lightness and spatial purity by concealing the 500mm-thick beam structures and equipment systems.

The roof is clad with graphite gray titanium zinc panels, a lightweight, corrosion-resistant material that suits the local humid climate. Its color complements the town's traditional gray roof tiles and white walls. Over time, the titanium zinc panels will develop a dense passivating protection layer, further enhancing the building's unique charm.

总平面图 / Site Plan

1	柳西半街	1	Liuxiban Street
2	展览馆	2	Exhibition Hall
3	青龙河	3	Qinglong River

张謇故里小镇位于清末民初著名实业家张謇的家乡南通，是绵延约4.7km²的乡村振兴暨古镇更新项目，旨在探索"产业+文旅"的乡村振兴模式。柳西半街位于小镇核心区，基于成于乾隆年间的常乐老街的整体更新而展开。设计对6处文保建筑和4处历史保护建筑予以修缮，保留青石板路等具有老街年代记忆的要素，并于此基础上植入商业及文化艺术功能，以尺度宜人的街巷、广场、庭园构建呼应当地历史传统的文化旅游体验。

老街东端新建一处展览馆，承担整座小镇文化地标的作用。展览馆场地北侧毗邻有着百年历史的木构保护建筑——常乐镇供销社。受当地民居原型"四汀宅沟"的启发，建筑师将建筑屋顶解构为四片简洁而优雅的单坡双曲面，化整为零的策略使得大体量新建筑与低矮老建筑相协调，巧妙化解了新旧建筑之间的尺度冲突。四片单坡双曲屋面向心汇聚后限定出窄院，天光云影浮动间，展览空间更显生动气韵。

展馆整体为钢结构，展厅内采用钢柱与墙体结合布置的方式，释放出内部宽敞的无柱空间。屋面构造被设计为端部收薄的"气枕"状空腔，500mm高的钢梁结构及设备系统隐藏于其中，保证了形式的轻盈感与空间的纯净感。

屋面选用石墨灰色钛锌板，这种材料具有轻质且天然抗腐蚀的性能，完美适配当地的湿润气候；石墨灰的色调也与镇区传统建筑的灰瓦白墙相得益彰。随着时间的推移，钛锌板将逐渐钝化形成致密保护层，为建筑增添独特的魅力。

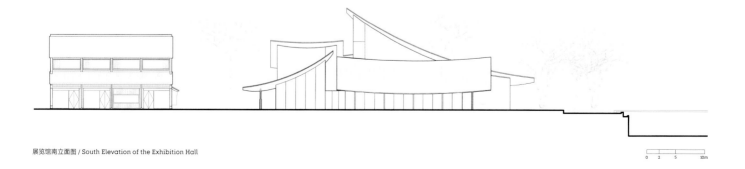

展览馆南立面图 / South Elevation of the Exhibition Hall

For

Public

公共设施

雅达剧院
Yada Theater

沪杭高速嘉兴服务区
G60 Expressway Jiaxing Service Area

建德市文化综合体
Jiande Cultural Center

舟山绿城育华幼儿园
Zhoushan Greentown Yuhua Kindergarten

飞鸟剧场
Earth Valley Theater

天台山雪乐园
Tiantaishan Snow Park

雅达剧院
Yada Theater

项目地点：江苏省无锡市
建筑面积：6,000m²
设计 / 竣工：2019/2022

Location: Wuxi, Jiangsu
Floor Area: 6,000m²
Design/Completion: 2019/2022

Yada Theater is a cultural icon of Yangxian Landscape in the Yangxian Ecological Tourism Area, surrounded by undulating mountains and lush bamboo forests. It delivers a new interpretation of the performance venue, fusing bodily experiences and eco aesthetics into its spatial organization. The theater embraces the pristine environment and regional spirit, bringing to community residents an open cultural parlor.

The pedestrian system follows the existing topography to preserve the mountainous landscape. The theater's boundaries are transformed into open courtyards interconnected by covered corridors, creating a free-flowing circulation. Limited built coverage generates an open, rhythmic ground space that encourages visitors to move between artificial structures and the natural environment.

The facade conveys Yixing's pottery and bamboo culture by employing locally fired ceramic panels and natural patterns. These panels exhibit a glazed celeste tone and delicate bamboo joint textures, which create a subtle color rhythm as they interact with shifting natural light and varying viewing distances.

A 530-seat multi-functional auditorium features a 21m-high floor-to-ceiling glass curtain wall. It can be converted into two configurations for concerts, dramas, or community cultural events. In the open stage mode, the landscape becomes the stage backdrop. By closing the curtain, the proscenium stage allows for customized stage settings and an immersive performance experience.

The Yada Theater adopts an integrated design approach, incorporating architecture, interior design, and landscape. Furthermore, specialized aspects such as lighting, acoustic systems, and curtain walls are cohesively coordinated to ensure a unified and holistic result.

总平面图 / Site Plan

1　喷泉　　1　Fountain
2　商业街　2　Commercial Street
3　会所　　3　Club

雅达剧院位于阳羡生态旅游度假区，是阳羡溪山小镇的重要文化地标。场地四周山峦起伏，竹林环绕。设计旨在探索开放式的观演空间类型，将身体经验与地貌特征融入空间秩序的构建策略之中，在延续场所的自然生态和人文特色的同时，为小镇居民带来一座开放的文化客厅。

剧院的内外边界被转译为一组对外开放的庭院序列，由局部风雨廊串联成一个自由的步行系统。动线设计顺应地貌，最大限度地保持山体的原始坡度。有限的覆盖创造出轻盈流动、节奏分明的底层空间，使人的身体不断穿梭于人工构筑与自然环境之间。

宜兴的陶、竹文化是本设计重要的灵感来源。雅达剧院的外立面采用当地烧制的釉面陶板为材料。天青色釉彩与竹节肌理板随着自然光线和观察距离的变化，呈现出微妙、动人的色彩韵律。

530座的多功能观演厅采用高21m的落地玻璃幕墙为主舞台背景，面向山水景观，可实现两种舞台模式的切换，不仅满足音乐会、戏剧演出的专业表演需求，也适用于多元的文化活动场景——透明模式将自然景色引入舞台，营造山林中观演的浪漫体验；镜框模式可落下幕布，满足戏剧演出定制化的布景需求。

本项目的建筑、室内、景观均为一体化设计。此外，照明、幕墙、声学等专项设计也由设计团队统一把控效果。

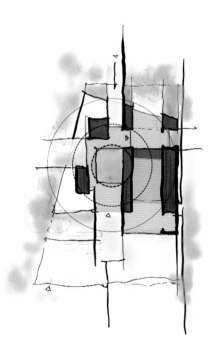

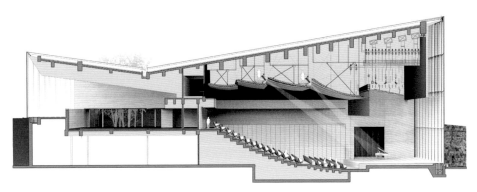

剖透视图 / Sectional Perspective

240　公共设施　　　For Public

沪杭高速嘉兴服务区
G60 Expressway Jiaxing Service Area

项目地点：浙江省嘉兴市
建筑面积：21,800m²
设计 / 竣工：2020/2022

Location: Jiaxing, Zhejiang
Floor Area: 21,800m²
Design/Completion: 2020/2022

The design began with rethinking the "expressway service area" typology. Challenging the conventional belief that these structures should solely serve functional purposes, the architects advocated for vibrant, multi-functional hubs that embrace and showcase diverse regional cultures.

Jiaxing Service Area displays the city's regional identity in a modern and symbolic manner within the highway landscape. It is separated into two distinct structures on the northern and southern sides of the expressway. The north building embodies a "Rippling Mirror," while the south represents a "Futuristic Vessel", together evoking the "water and ship" symbolism of Jiaxing culture.

This asymmetrical design departs from conventional approaches and assists travelers in identifying their directions. The facade of the south building exudes a futuristic aesthetic with its three-dimensional modular stainless-steel sheets. In contrast, the north building showcases undulating polycarbonate sheets on the exterior and concrete panels on the open ground floor ceiling, creating a design reminiscent of the waves on South Lake.

The design provides users a fresh and captivating journey by redefining elements such as the parking experience, public scenarios, and travel memories. It raises the main functional spaces to the second and third floors while utilizing the ground floor for parking, maximizing the use of the limited land area. In the humid subtropical climate of Jiangnan, this sheltered ground floor provides a more comfortable environment for drivers and passengers.

The entrance lobby is centrally located on the open ground level and complemented by a three-story atrium featuring sleek escalators, including a direct connection to the third floor. This design caters to various travel scenarios and time constraints, offering travelers convenient options. Those in a rush can quickly access the restrooms, restaurants, and ground-floor exit using the cross-floor escalator.

本项目设计的起点是对"高速服务区"这一特定建筑类型的审视与再思考。建筑师认为，当代服务区不应被局限为仅提供基本服务的功能性构筑物，而应被视作充满活力的复合型服务场所、地域文化的展示平台。

服务区位于广阔平原之上，设计选择以写意的方式于公路场景中传达嘉兴的地域特征。以"未来之舟""南湖映波"为主题，南北两侧综合楼以方盒子的形式漂浮于公路之上，诠释来自嘉兴文化的"依水行舟"意象。

南北综合楼差异化的材料与造型手法有利于方向的区分，是区别于传统服务区的一次尝试。北区立面运用波浪形聚碳酸酯板，架空层天花采用波浪形水泥板，曲面肌理暗示嘉兴南湖水波荡漾的美景；南区则运用三维造型的单元式不锈钢板传达未来感。

设计从停车方式、活动场景、旅途印象等各个方面入手，重新定义了服务区的体验。在有限的用地中，建筑师大胆采用了首层架空的设计策略，将主要使用功能抬升至二、三层，使地面能够最大限度地为停车所用。基于江南地区多雨、夏季炎热的气候特征，架空层的设计大大提升了司乘人员的使用体验。

服务区入口门厅设置在架空层的中央位置，三层通高的中庭布置有醒目的垂直交通设施，包括一架直通三层的飞天梯——这样的设置为具有不同时间诉求的旅客预留了行为可能性，匆忙者也可方便地通过跨层扶梯直达洗手间或简餐处并快速离开。

1	商业	1	Commercial
2	停车区	2	Parking

北区剖透视图 / North Building Sectional Perspective

建德市文化综合体
Jiande Cultural Center

项目地点：浙江省杭州市
建筑面积：44,900m²
设计/竣工：2013/2021

Location: Hangzhou, Zhejiang
Floor Area: 44,900m²
Design/Completion: 2013/2021

Jiande Cultural Center is a significant urban landmark. It sits in the eastern heart of the city, nestled between the newly developed Qiaodong business district in the north and a picturesque landscape of rolling hills along the Xin'an River in the south.

The site is elongated, spanning 600m from east to west and ranging in depth from 60-140m from north to south. It is situated along the riverbanks, facing an approximately 280m-wide water surface. With an above-ground floor area of approximately 31,900m², this municipal cultural complex comprises four facilities: a library, a museum, a women's and children's activity center, and a youth activity center.

The water of the Xin'an River originates from an upstream reservoir and maintains a stable temperature year-round. As a result of seasonal temperature fluctuations, the river's surface is often shrouded in dense fog. Reflecting the aesthetic essence of the misty landscape, the design embraces the ethereal qualities of the fog as its guiding inspiration. The main functional spaces are elevated to the 2nd and 3rd floors, while the ground floor is released, creating a connection between the pedestrian circulation, urban roads, and the waterfront landscape. The entire structure resembles a winding ribbon that gracefully "floats" above the site. To align with the urban context, the northern interface breaks down into smaller individual components, blending with the scale of the surrounding buildings.

The ribbon-shaped canopy roof provides pedestrians with a "sheltering wing" and creates an inviting space for various activities. Its 40m arc span forms a striking frame where the central plaza intersects the urban axis. To enhance the visual relationship between the city and the natural surroundings, the span structures are minimized to four 14m-tall steel columns.

The traditional perception of cultural facilities as solemn, sublime, and dignified has evolved into a more inclusive and accessible communal platform. The Jiande Cultural Center responds to this shift by offering transparent and interconnected spaces, embodying a cultural landmark that is light, elegant, gentle, and appealing.

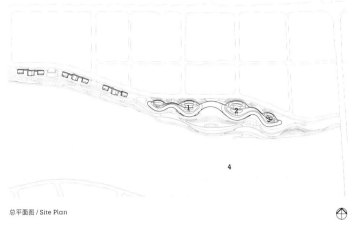

总平面图 / Site Plan

1	图书馆	1 Library
2	博物馆	2 Museum
3	活动中心	3 Activity Center
4	新安江	4 Xin'an River

建德市文化综合体是杭州建德重要的地标建筑和城市名片。基地位于建德桥东区域核心地段，南临新安江，北侧为桥东中心商务区。

场地狭长，东西长约600m，南北进深在60m至140m之间，所临的新安江水面宽约280m。作为一座市级文化综合体，建筑将容纳图书馆、博物馆、妇女儿童活动中心、青少年活动中心四个市民空间，地上建筑面积共约31,900m²。

受新安江上游水库影响，基地毗邻的江面上常能形成"白沙奇雾"的景观。受此启发，建筑以"悬浮"的姿态回应缥缈的山水意境。主要功能空间被抬高至二、三层，底层空间则被释放出来，与城市街道、滨水空间完全连通。整座文化综合体如同一条蜿蜒的飘带，舒展、低平地飘浮于场地之上。北侧界面化整为零，以较小体量营造怡人的尺度。

悬浮的屋顶也为滨江带的行人带来一片"庇护的羽翼"，为多样的活动带来可能性。屋檐中段的广场衔接城市轴线与滨江的水秀广场，40m的大跨屋面由四根14m高的钢柱承托，勾勒出城市与山水之间的抽象景框。

今日的文化设施建筑已经从严肃崇高、神秘厚重的传统形象逐步转变为市民文化生活的服务平台。建德市文化综合体凭借通透连续的空间实现了这一诉求，同时也为城市展示出一座文化地标轻盈、典雅、柔和而稳重的美好形象。

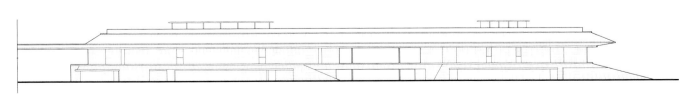

南立面图 / South Elevation

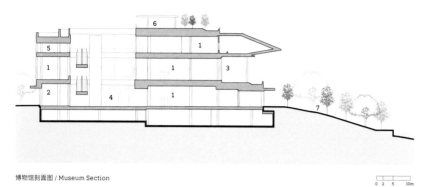

博物馆剖面图 / Museum Section

1	展厅	1 Exhibition Hall
2	大厅	2 Lobby
3	公共平台	3 Public Platform
4	中庭	4 Atrium
5	办公室	5 Office
6	屋顶花园	6 Roof Garden
7	滨江景观带	7 Riverside Landscape Belt

舟山绿城育华幼儿园
Zhoushan Greentown Yuhua Kindergarten

项目地点：浙江省舟山市
建筑面积：13,500m²
设计/竣工：2018/2021

Location: Zhoushan, Zhejiang
Floor Area: 13,500m²
Design/Completion: 2018/2021

The design concept of Zhoushan Greentown Yuhua Kindergarten arose from a consensus among the architects and the teaching team to establish an environment where children can actively engage in learning rather than passively receive predetermined educational objectives. In this way, the kindergarten serves as the children's "third instructor."

The master plan adopts a courtyard layout, with 18 classrooms arranged along the eastern and southern site edges, creating a gentle curve that embraces a small central playground.

Based on this layout, the design establishes a network of interconnected public spaces to ensure ample spatial freedom and foster active exploration. Special classrooms, such as the crafting room, marine room, baking room, and drama room, are strategically placed at the intersection of corridors.

Different-level terraces connect indoor and outdoor areas, offering ample outdoor activity space on each floor. The winding circular pedestrian system unifies the entire structure, creating a three-dimensional, playful circulation that efficiently links various functional spaces. These pathways lead to the rooftop, where children can enjoy the entire cityscape.

The kindergarten adopts an integrated architecture and landscape design to create visually captivating, nature-oriented spaces. "Transparent" interfaces between indoor and outdoor areas enable children to enjoy courtyard landscapes from a lower vantage point, promoting a sense of spatial openness and facilitating teacher supervision. Natural light is introduced into the public corners, guiding children to perceive spatial transitions and time passages in the environment.

舟山绿城育华幼儿园的设计始于建筑师与教研团队的共识：幼儿园的环境应成为"孩子的第三位老师"，通过支持幼儿的探索，引领幼儿成为积极的知识构造者，而非成为被动的教学目标。

本项目采用围合式布局，18个教室围绕中央的小操场沿基地东、南两侧展开。在此基础之上，建筑师构建了一个多路径交织的公共空间系统，以充分的空间自由度鼓励儿童的自主探索。特殊功能教室如建构室、海洋室、烘焙室、戏剧室等均分布于路径的交汇处。

室内外空间利用不同标高的平台相互连接，为各层提供充足的露天活动场地。蜿蜒的环形流线串联整栋建筑，既建立了三维立体的嬉戏路径，又高效地连接了各功能空间。路径一直通往屋顶，孩子们可以在这里眺望城市。

幼儿园通过建筑景观一体化设计，构筑视野丰富、亲近自然的活动场所。室内、外界面的透明化处理，确保孩子们能够以较低的视点观赏到开阔的庭院景观，同时也提升了教师照管看护的效率。公共区域引入自然光线，提示空间的转换和时间的变化，引导儿童用身体感受环境中蕴含的丰沛信息。

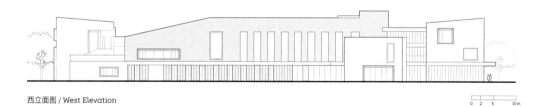

西立面图 / West Elevation

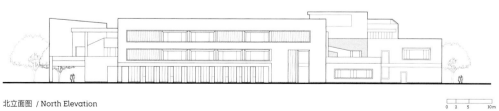

北立面图 / North Elevation

253

飞鸟剧场
Earth Valley Theater

项目地点：江苏省无锡市
建筑面积：9,200m²
设计/竣工：2021/—
Location: Wuxi, Jiangsu
Floor Area: 9,200m²
Design/Completion: 2021/-

256　公共设施　　For Public

Earth Valley Theater is the first Chinese open-air performance venue with the theme of human-bird interactive experiences. Situated within the Mingling Mountains to the southwest of "Pottery Capital" Yixing City, it is part of Yaohu Town. The design revolves around the idea of "an invitation from birds," awakening an entrancing avian kingdom.

Stretching east to west, a slender valley converges with an open expanse at the site's western edge. With a 270m-deep valley and a 240m-wide open area, the site is hugged by luxuriant bamboo forests on both sides of the mountains. Appreciating this gorgeous natural setting, the architect imagines the theater as a meandering piece of earth art, with each step as a prelude to the performance.

The theater contains three distinctive sections: a central auditorium that can accommodate 2,300 individuals; a terrace area for ticketing services, retail outlets, and logistical support; and an aviary zone that doubles as a bird observation platform. The design brings new life to locally-sourced "clay" and "bamboo" materials. The auditorium and terrace area are themed by "clay." The combination of terracotta, bricks, and tiles, all crafted from sculpted concrete, brings a dynamic texture and integrates the structure with the earth. The aviaries are enveloped in intricately woven bamboo-textured sheets, serving as a stage backdrop that connects to the forest.

In this project, GOA facilitated an integrated design approach incorporating landscape, architecture, and interior by collaborating with professionals from various fields. This collaboration integrates Director Fan Yue's artistic vision with GOA's architectural philosophy, forming a synergistic relationship where each concept enhances the other. The avian consultant team provided meticulous technical support in the intricate design of the aviary zone and performance area. Additionally, the material consultant team, renowned for its expertise in Disney landscape design, has contributed significantly to implementing and applying materials.

总平面图 / Site Plan

1	鸟舍区	1	Aviary
2	入口展厅	2	Exhibition Hall
3	舞台	3	Stage
4	观众席	4	Amphitheater
5	叠台区	5	Terrace

剧场位于陶都宜兴西南的茗岭山山谷之中，是窑湖小镇的一部分，也是国内首座以人鸟互动为主题的日间露天演出场所。以"鸟的邀约"为主题，剧场旨在呈现一个自然之中的鸟的王国。

项目基地由一条东西向的狭谷和西侧开阔地带组成，沿峡谷进深约270m，开阔处宽约240m，周边山体为竹林所覆盖。受张力十足的自然环境启发，建筑师将剧场视作一件葡萄于山谷之中的大地艺术作品——进入剧场，便意味着表演的开始。

剧场空间分为三部分：可容纳2300人的观演区；包含剧场后勤、商业、售票等功能的叠台区；包含鸟类饲养和活动空间、兼作鸟类观赏之用的鸟舍区。在地材料"陶"与"竹"在建筑中获得新的生命力。观演区与叠台区以"陶"为材料主题，由雕刻混凝土仿陶土材料制成的陶土、陶泥、陶片在建筑与景观之间交替变幻，使得建筑与大地浑然一体。鸟舍区对"竹"进行再演绎，仿竹篾编织单元组成的自由曲面包裹着鸟舍单元，作为舞台的背景融于后方的竹海之中。

本项目中，goa大象设计作为建筑、景观、室内一体化设计服务方，与多元领域的专业团队展开跨界合作。在设计的推进过程中，樊跃导演团队的演出设计创意与建筑设计方案双线并行、相辅相成；鸟类顾问团队为鸟舍区及表演区的设计细节提供了详尽的技术支持；参与迪士尼景观设计的材料顾问团队深度介入了"陶"与"竹"的实现与落地。

观演区剖面图 / Amphitheater Section

1 商业　　1 Retail
2 观众席　2 Auditorium
3 舞台　　3 Stage

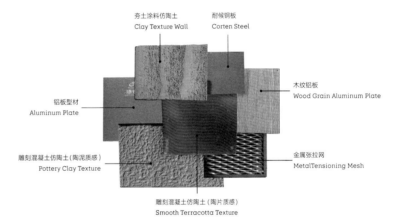

- 夯土涂料仿陶土 Clay Texture Wall
- 耐候钢板 Corten Steel
- 木纹铝板 Wood Grain Aluminum Plate
- 铝板型材 Aluminum Plate
- 雕刻混凝土仿陶土（陶泥质感）Pottery Clay Texture
- 金属张拉网 MetalTensioning Mesh
- 雕刻混凝土仿陶土（陶片质感）Smooth Terracotta Texture

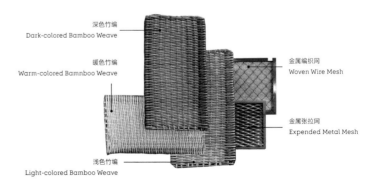

- 深色竹编 Dark-colored Bamboo Weave
- 暖色竹编 Warm-colored Bamnboo Weave
- 金属编织网 Woven Wire Mesh
- 金属张拉网 Expended Metal Mesh
- 浅色竹编 Light-colored Bamboo Weave

天台山雪乐园
Tiantaishan Snow Park

所在地址：浙江省台州市
建筑面积：19,400m²
设计/竣工：2018/2021
Location: Taizhou, Zhejiang
Floor Area: 19,400m²
Design/Completion: 2018/2021

The Tiantaishan Snow Park is at an altitude of 800m within Tiantai Mountain, a 5A National Tourist Attraction celebrated for its religious significance and breathtaking landscapes. As the first winter sports-themed amusement park in Taizhou City, this project combines sports activities and cultural tourism within an appealing architectural design.

The facade's design abstracts the form of a lotus, a symbol of Buddhist culture, into a geometric pattern, adding a touch of purity to the architectural expression within the tranquil mountain landscape. This folding envelope utilizes white perforated and solid aluminum panels to generate a dynamic play of light and shadow that changes with the light. Each facade element features a staggered "hexagon" texture reminiscent of delicate snowflakes.

Adhering to a site-specific adaptation principle, the ski routes are planned to align with the mountain's natural contours, minimizing the need for excavation, construction, and associated costs. A sequence of expansive outdoor platforms connects the structure and exterior roads. These platforms, with their curves resembling a gracefully wrapped ribbon, mitigate the visual impact of the architecture and offer multiple vantage points to appreciate the scenic beauty while enhancing the lower-level spatial experience.

The indoor ski area spans approximately 15,000m² and is divided into three main functional zones: the service, the ski, and the snow entertainment. The cold zone comprises about two-thirds of the total area and includes standard ski slopes, recreational facilities, a youth play area, a landscape garden, and an ice theater. Along the boundary between the cold and warm zones are expansive floor-to-ceiling windows and cozy tea chambers, inviting guests to relax and offering views of the skiing and entertainment areas.

天台山雪乐园坐落在以"佛宗道源、山水神秀"享誉海内外的国家5A级风景名胜区天台山，地处海拔800m的山峦腹地。该项目是台州市首个综合性冰雪主题乐园，也是一座融"运动与文旅"于一体的特色建筑。

立面灵感来自"莲"的意象，"看取莲花净，应知不染心"——莲花作为佛教文化元素，与天台山淡雅高远的山景意境相谐。建筑师抽象转译莲的形态，以白色穿孔铝板和实面铝板拼接形成几何折面。穿孔板图案是犬牙交错的六边形，呼应雪花之形。在不同光线下，莲瓣呈现出丰富多变的明暗光影效果。

建筑利用山体自然坡度布置雪道，有效减少了土方工程量及建造成本。一组宽大的室外平台架设于建筑与山体之间，起到衔接外部道路的作用。曲线形式的平台如同套在场馆体量上的蜿蜒绸带，既柔化了场馆的厚重体量，又提供了多角度的观山视角，为底层带来丰富的灰空间。

室内滑雪场建筑面积约15,000m²，分为三大功能区：服务区、滑雪区、娱雪区；其中冷区面积占比约2/3，不仅囊括基础滑雪场和冰雪娱乐设施，还设有儿童游戏区、景观区、表演区等。冷区与暖区的分界处设有超大落地观景玻璃窗及可鸟瞰滑雪区、娱雪区的休闲茶座。

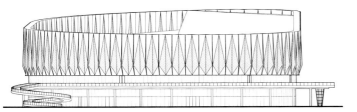

北立面图 / North Elevation

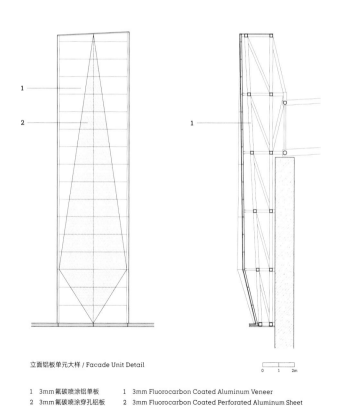

立面铝板单元大样 / Facade Unit Detail

1　3mm 氟碳喷涂铝单板　　1　3mm Fluorocarbon Coated Aluminum Veneer
2　3mm 氟碳喷涂穿孔铝板　　2　3mm Fluorocarbon Coated Perforated Aluminum Sheet

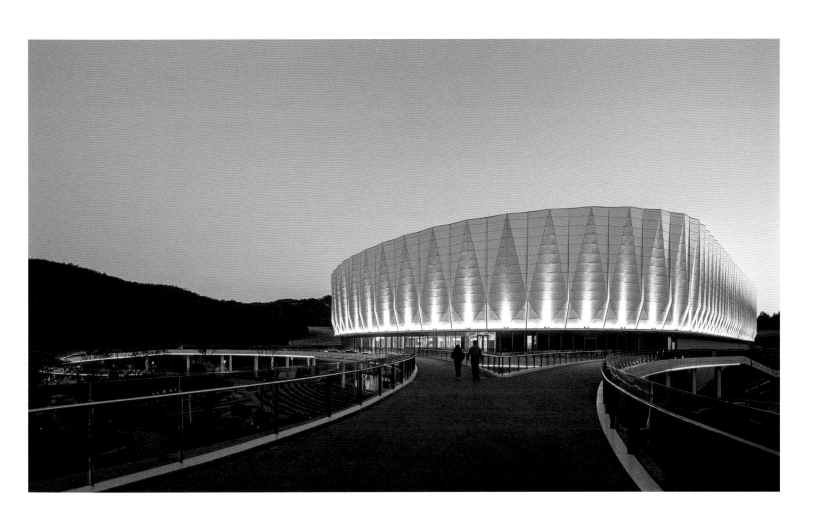

核心团队
Core Team

陆 皓
LU Hao

总裁 / 总建筑师 / 资深合伙人
President / Senior Principal

国家一级注册建筑师
1RA-PRC
英国皇家特许注册建筑师
RIBA Chartered Architect
世界高层建筑与都市人居学会特邀专家
CTBUH Designated Expert

何 兼
HE Jian

执行总裁 / 总建筑师 / 资深合伙人
Executive President / Senior Principal

正高级工程师 / 国家一级注册建筑师
Professor-level Senior Engineer / 1RA-PRC
英国皇家特许注册建筑师 / 上海市建筑学会理事
RIBA Chartered Architect / Director of ASSC
世界高层建筑与都市人居学会特邀专家
CTBUH Designated Expert

何 峻
HE Jun

执行总裁 / 总建筑师 / 资深合伙人
Executive President / Senior Principal

国家一级注册建筑师
1RA-PRC

凌 建
LING Jian

执行总裁 / 总建筑师 / 资深合伙人
Executive President / Senior Principal

国家一级注册建筑师
1RA-PRC
世界高层建筑与都市人居学会特邀专家
CTBUH Designated Expert

李 慧芬
LI Huifen

总经理 / 资深合伙人
General Manager / Senior Principal

设计
Architect

汪 澜
WANG Lan

总建筑师 / 资深合伙人
Senior Principal

国家一级注册建筑师
1RA-PRC

荣 嵘
RONG Rong

总建筑师 / 资深合伙人
Senior Principal

国家一级注册建筑师
1RA-PRC

田 钰
TIAN Yu

总建筑师 / 资深合伙人
Senior Principal

张 晓晓
Shawn CHEUNG

总建筑师 / 资深合伙人
Senior Principal

国家一级注册建筑师
1RA-PRC
英国皇家特许注册建筑师
RIBA Chartered Architect

孙 航
SUN Hang

总建筑师 / 资深合伙人
Senior Principal

国家一级注册建筑师
1RA-PRC

陈 斌鑫
CHEN Binxin

总建筑师 / 资深合伙人
Senior Principal

国家一级注册建筑师
1RA-PRC
世界高层建筑与都市人居学会特邀专家
CTBUH Designated Expert

袁 源
YUAN Yuan

总建筑师 / 资深合伙人
Senior Principal

国家一级注册建筑师
1RA-PRC

张 迅
ZHANG Xun

总建筑师 / 资深合伙人
Senior Principal

国家一级注册建筑师
1RA-PRC

王 彦
WANG Yan

总建筑师 / 合伙人
Principal

英国皇家特许注册建筑师
RIBA Chartered Architect
瑞士工程师和建筑师协会会员
SIA Member
同济大学建筑与城市规划学院客座教授
Visiting Prof. CAUP Tongji University

陈 健
CHEN Jian

总建筑师 / 合伙人
Principal

梁 卓敏
LIANG Zhuomin
总建筑师 / 合伙人
Principal

国家一级注册建筑师
1RA-PRC

韩 中强
HAN Zhongqiang
总建筑师 / 合伙人
Principal

国家一级注册建筑师
1RA-PRC

严 育林
YAN Yulin
总建筑师 / 合伙人
Principal

国家一级注册建筑师
1RA-PRC

徐 琦
XU Qi
上海办公室总经理 / 总建筑师 / 合伙人
General Manager of GOA Shanghai /
Principal

国家一级注册建筑师
1RA-PRC

张 琪琳
ZHANG Qilin
总建筑师 / 合伙人
Principal

国家一级注册建筑师
1RA-PRC

徐 匀飞
XU Yunfei
副总建筑师
Director

国家一级注册建筑师
1RA-PRC

李 震
LI Zhen
副总建筑师
Director

国家一级注册建筑师
1RA-PRC

侯 冬炜
HOU Dongwei
副总建筑师
Director

陈 吉
CHEN Ji
副总建筑师
Director

国家一级注册建筑师
1RA-PRC

周 翌
ZHOU Yi
副总建筑师
Director

国家一级注册建筑师
1RA-PRC

陈 伟
CHEN Wei
副总建筑师
Director

国家一级注册建筑师
1RA-PRC

陈 清西
CHEN Qingxi
设计总监
Director

国家一级注册建筑师
1RA-PRC

蒋 嘉菲
JIANG Jiafei
设计总监
Director

国家一级注册建筑师
1RA-PRC

于 军贵
YU Jungui
设计总监
Director

胡 晨芳
HU Chenfang
设计总监
Director

国家一级注册建筑师
1RA-PRC

陈 娴
CHEN Xian
设计总监
Director

国家一级注册建筑师
1RA-PRC

刘 波
LIU Bo

设计总监
Director

国家一级注册建筑师
1RA-PRC

袁 波
YUAN Bo

设计总监
Director

吕 东旭
LYU Dongxu

设计总监
Director

吕 焕政
LYU Huanzheng

设计总监
Director

李 洪
LI Hong

设计总监
Director

国家一级注册建筑师
1RA-PRC

韦 栋安
WEI Dong'an

设计总监
Director

国家一级注册建筑师
1RA-PRC

何 松
HE Song

设计总监
Director

郑 文康
ZHENG Wenkang

设计总监
Director

国家一级注册建筑师
1RA-PRC

徐 正
XU Zheng

设计总监
Director

李 凌
LI Ling

设计总监
Director

国家一级注册建筑师
1RA-PRC

技术
Engineer

黄 伟志
HUANG Weizhi

副总工程师/合伙人
Principal

国家一级注册结构工程师
1RSE-PRC

杜 立明
DU Liming

总建筑师
Director

国家一级注册建筑师
1RA-PRC

寿 广
SHOU Guang

总工程师
Director (MEP)

正高级工程师
Professor-level Senior Engineer
注册公用设备工程师（暖通空调）
RUE (HVAV)

师 建伟
SHI Jianwei

副总工程师
Director (STRU)

国家一级注册结构工程师
1RSE-PRC

李 令捷
LI Lingjie

技术总监
Director

国家一级注册建筑师
1RA-PRC

管理
Manager

徐 浩祥
XU Haoxiang

技术总监
Director (STRU)

国家一级注册结构工程师
1RSE-PRC

郭 吟
GUO Yin

技术总监
Director

国家一级注册建筑师
1RA-PRC

陈 勇
CHEN Yong

技术总监
Director

方 旭筠
FANG Xuyun

资深合伙人
Senior Principal

刘 纲
LIU Gang

上海办公室总裁 / 总建筑师 / 资深合伙人
President of GOA Shanghai / Senior Principal

国家一级注册建筑师
1RA-PRC
世界华人建筑师协会理事
Director of WACA

卿 州
QING Zhou

执行总经理
Executive General Manager

英国皇家特许注册建筑师
RIBA Chartered Architect
世界高层建筑与都市人居学会特邀专家
CTBUH Designated Expert

董 琰
DONG Yan

上海办公室执行总经理
Executive General Manager of GOA Shanghai

包 风
BAO Feng

高级信息总监
Information Director

国家一级注册结构工程师
1RSE-PRC
注册土木工程师（岩土）
RCE (Geotechnical)

王 旭炯
WANG Xujiong

高级运营总监
Director (OP.)

徐 碧霞
XU Bixia

运营总监
Director (OP.)

国家一级注册建筑师
1RA-PRC

张 孟东
ZHANG Mengdong

财务总监
Finance Director

沈 旭凯
SHEN Xukai

运营总监
Director (OP.)

正高级工程师
Professor-level Senior Engineer
国家一级注册结构工程师
1RSE-PRC

杨 婉娜
YANG Wanna

人力资源总监
Human Resources Director

沈 超
SHEN Chao

运营总监
Director (OP.)

沈 毅
SHEN Yi

运营总监
Director (OP.)

项目资料
Project Data

goa 大象设计总部
GOA Headquarters

业主: goa 大象设计
委托方式: 直接委托
建设需求: 新建
项目地点: 浙江省杭州市
建筑面积: 16,900m²
设计 / 竣工: 2018/2020
设计团队:
项目负责: 陆皓
建筑设计师: 杜立明、卿州、刘琳
室内设计师: 林琳赟、诸双、戴亮亮、胡文涛、郭思聪、贾一帆、袁佳琛
结构工程师: 胡凌华、包风、龚铭、杨洋、詹伟良、庞振钱
设备工程师: 王文胜、周益林、黄池钧、葛健、花勇刚、葛令科、孙中南、梅玉龙、寿广、王超伟、程磊、庄少阳、叶金元、何刚、张雪祁、赵志铭、张毓、钱列东、徐幸、刘佳莹
合作单位:
标识顾问: 702 Design
摄影: goa 大象设计

Client: Group of Architects
Commission: Client Brief
Construction Brief: New Construction
Location: Hangzhou, Zhejiang
Floor Area: 16,900m²
Design/Completion: 2018/2020
Design Team:
Lead Architects: LU Hao
Architecture: DU Liming, QING Zhou, LIU Lin
Interior: LIN Linyun, ZHU Shuang, DAI Liangliang, HU Wentao, GUO Sicong, JIA Yifan, YUAN Jiachen
Structure: HU Linghua, BAO Feng, GONG Ming, YANG Yang, ZHAN Weiliang, PANG Zhenqian
MEP: WANG Wensheng, ZHOU Yilin, HUANG Chijun, GE Jian, HUA Yonggang, GE Lingke, SUN Zhongnan, MEI Yulong, SHOU Guang, WANG Chaowei, CHENG Lei, ZHUANG Shaoyang, YE Jinyuan, HE Gang, ZHANG Xueqi, ZHAO Zhiming, ZHANG Yu, QIAN Liedong, XU Xing, LIU Jiaying
Collaborators:
Signage: 702 Design
Images: GOA

上海西岸金融城 G 地块
Shanghai West Bund Financial City - Plot G

业主: 香港置地
委托方式: 直接委托
建设需求: 新建 + 改造
项目地点: 上海市徐汇区
建筑面积: 97,500m²
设计 / 竣工: 2020/—
设计团队:
项目负责: 陆皓、张迅、徐琦
建筑设计师: 窦志国、胡培、黄里达、花子杰、柴敬、沈强、李力、陈振、张小蓉、孙照人、徐东波、林发光、张增鑫、江皓、何广竣
结构工程师: 严志威
设备工程师: 任庆军、葛令科、徐幸、南旭、陈文卉
合作单位:
室内设计: HWCD
景观设计: 翡世景观
施工图设计: 天华
历史建筑保护顾问: 上海章明建筑设计事务所
摄影: 陈曦工作室

Client: HongKong Land
Commission: Client Brief
Construction Brief: New Construction & Renovation
Location: Xuhui, Shanghai
Floor Area: 97,500m²
Design/Completion: 2020/-
Design Team:
Lead Architects: LU Hao, ZHANG Xun, XU Qi
Architecture: DOU Zhiguo, HU Pei, HUANG Lida, HUA Zijie, CHAI Jing, SHEN Qiang, LI Li, CHEN Zhen, ZHANG Xiaorong, SUN Zhaoren, XU Dongbo, LIN Faguang, ZHANG Zengxin, JIANG Hao, HE Guangjun
Structure: YAN Zhiwei
MEP: REN Qingjun, GE Lingke, XU Xinq, NAN Xu, CHEN Wenhui
Collaborators:
Interior: Harmony World Consultant and Design
Landscape: FISH DESIGN
Construction Drawing: TIANHUA
Historic Preservation Consultant: Shanghai Zhang Ming Architectural Design Firm
Images: CHEN Xi Studio

鹿山时代
Lushan Times

业主: 山水置业
委托方式: 直接委托
建设需求: 新建
项目地点: 浙江省杭州市
建筑面积: 173,000m²
设计 / 竣工: 2013/2020
设计团队:
项目负责: 何兼、袁源、韩中强
建筑设计师: 汪海莹、叶帆、顾江昱、厉孜浣、胡佳佳、潘强强、向培
结构工程师: 师建伟、金晓东、龚铭、沈旭凯、金明彦、王申昊、敖国胜
设备工程师: 寿广、叶金元、程磊、王超伟、江漪波、吴文裕、胡一东、王雅迪、钱列东、吴金祥、郑铭、陈梦洁、李程、黄建军、梅玉龙、杨福华、毛迪华、平俊晖、花勇刚、黄路、王晓卉
合作单位:
室内设计: 品冠设计
景观设计: 伍道国际
摄影: 清筑影像

Client: Shanshui Real Estate
Commission: Client Brief
Construction Brief: New Construction
Location: Hangzhou, Zhejiang
Floor Area: 173,000m²
Design/Completion: 2013/2020
Design Team:
Lead Architects: HE Jian, YUAN Yuan, HAN Zhongqiang
Architecture: WANG Haiying, YE Fan, GU Jiangchan, LI Zihuan, HU Jiajia, PAN Qiangqiang, XIANG Pei
Structure: SHI Jianwei, JIN Xiaodong, GONG Ming, SHEN Xukai, JIN Mingyan, WANG Shenhao, AO Guosheng
MEP: SHOU Guang, YE Jinyuan, CHENG Lei, WANG Chaowei, JIANG Yibo, WU Wenyu, HU Yidong, WANG Yadi, QIAN Liedong, WU Jinxiang, ZHENG Ming, CHEN Mengjie, LI Cheng, HUANG Jianbao, MEI Yulong, YANG Fuhua, MAO Dihua, PING Junhui, HUA Yonggang, HUANG Lu, WANG Xiaohui
Collaborators:
Interior: Pinguan Design
Landscape: Wonderway International
Images: CreatAR

深圳万创云汇 01-03 地块
Shenzhen Vanke Cloud Gradus - Plot 01-03

业主: 万科集团
委托方式: 直接委托
建设需求: 新建
项目地点: 广东省深圳市
建筑面积: 224,900m²
设计 / 竣工: 2021/—
设计团队:
项目负责: 张迅
建筑设计师: 董清源、柴敬、朱佳辉、顾方荣、宋梦娇、何广竣、于之淳
结构工程师: 杨洋、王申昊、赵凯龙
设备工程师: 张雪祁、周伟明、杨福华
合作单位:
室内设计: HBA
景观设计: ALN
施工图设计: 广东省建筑设计研究院深圳分院

Client: Vanke
Commission: Client Brief
Construction Brief: New construction
Location: Shenzhen, Guangdong
Floor Area: 224,900m²
Design/Completion: 2021/-
Design Team:
Lead Architect: ZHANG Xun
Architecture: DONG Qingyuan, CHAI Jing, ZHU Jiahui, GU Fangrong, SONG Mengjiao, HE Guangjun, YU Zhichun
Structure: YANG Yang, WANG Shenhao, ZHAO Kailong
MEP: ZHANG Xueqi, ZHOU Weiming, YANG Fuhua
Collaborators:
Interior: Hirsch Bedner Associates
Landscape: Adrian L. Norman Ltd.
Construction Drawing: GDAD Shenzhen Branch

上海长风中心
Shanghai Changfeng Center

业主: 上海浙铁绿城房地产开发有限公司
委托方式: 直接委托
建设需求: 新建
项目地点: 上海市普陀区
建筑面积: 359,800m²
设计 / 竣工: 2011/2021
设计团队:
项目负责: 凌建
建筑设计师: 陈吉、陈清西、杜立明、宓芬、钱炎辉、兰小明、楼澎箐、李海明、胥小斌、闵立然、刘大可
结构工程师: 胡凌华、包风、何亮、柴磊、詹伟良、杜攀峰、崔碧琪、吴玄成、金明彦
设备工程师: 叶金元、庄少阳、朱丹、邓雅琼、寿广、周伟明、钱列东、郑铭、侯会芬、陈梁星、黄腾、黄建宝、张宇明、梅玉龙、孙中南、王文胜、黄路、陈舟舟、纪殿格、毛迪华、李翔、张旭、廖国勇、王晓卉、许光红
合作单位:
室内设计: 伍兹贝格
景观设计: 艾奕康
摄影: goa 大象设计

Client: Shanghai Zhetie Greentown Real Estate Development Co., Ltd.
Commission: Client Brief
Construction Brief: New Construction
Location: Putuo, Shanghai
Floor Area: 359,800m²
Design/Completion: 2011/2021
Design Team:
Lead Architect: LING Jian
Architecture: CHEN Ji, CHEN Qingxi, DU Liming, MI Feng, QIAN Yanhui, LAN Xiaoming, LOU Hanjing, LI Haiming, XU Xiaobin, MIN Liran, LIU Dake
Structure: HU Linghua, BAO Feng, HE Liang, CHAI Lei, ZHAN Weiliang, DU Panfeng, CUI Biqi, WU Xuancheng, JIN Mingyan
MEP: YE Jinyuan, ZHUANG Shaoyang, ZHU Dan, DENG Yaqiong, SHOU Guang, ZHOU Weiming, QIAN Liedong, ZHENG Ming, HOU Huifen, CHEN Liangxing, HUANG Teng, HUANG Jianbao, ZHANG Yuming, MEI Yulong, SUN Zhongnan, WANG Wensheng, HUANG Lu, CHEN Zhouzhou, JI Diange, MAO Dihua, LI Xiang, ZHANG Xu, LIAO Guoyong, WANG Xiaohui, XU Guanghong
Collaborators:
Interior: Woods Bagot
Landscape: AECOM
Images: GOA

重庆启元
Century Land Chongqing

业主: 香港置地
委托方式: 直接委托
建设需求: 新建
项目地点: 重庆市江北区
建筑面积: 175,400m²
设计 / 竣工: 2020/—
设计团队:
项目负责: 陆皓、张迅、陈健
建筑设计师: 胡培、涂兆云、张腾撼、秦闻怡、耿琳、郭绣沅津、杨淑婷、徐成浩、王骏戈
结构工程师: 俞洪
设备工程师: 吴金祥、高利强、吴文裕
合作单位:
室内设计: DPH 设计事务所、矩阵纵横、GBD 杜文彪设计
景观设计: 澳派景观设计工作室（商业）、加特林景观（住宅）
施工图设计: 中机中联（商业）、中泰设计（住宅）
摄影: 存在建筑、香港置地

Client: Hongkong Land
Commission: Client Brief
Construction Brief: New Construction
Location: Jiangbei, Chongqing
Floor Area: 175,400m²
Design/Completion: 2020/-
Design Team:
Lead Architect: LU Hao, ZHANG Xun, CHEN Jian
Architecture: HU Pei, TU Zhaoyun, ZHANG Tenghan, QIN Tianyi, GENG Lin, GUO Mianyuanjin, YANG Shuting, XU Chenghao, WANG Junge
Structure: YU Hong
MEP: WU Jinxiang, GAO Liqiang, WU Wenyu
Collaborators:
Interior: Design Power House, Matrix Design, GBD
Landscape: ASPECT Studios (commercial), JTL Studio (residential)
Construction Drawing: CMCU (commercial), ZTA (residential)
Images : Arch-Exist, Hongkong Land

恒力环企中心
Hengli Global Enterprise Center

业主：恒力集团
委托方式：招投标
建设需求：新建
项目地点：江苏省苏州市
建筑面积：1,277,000m²
设计 / 竣工：2020/—
设计团队：
项目负责：凌建、田钰、韩中强
建筑设计师：郭吟、张弢、王勇、寇广建、俞翔、蒋寅、叶帆、赵灵燕、邵笑琦、向陪、郑珏、贾彦琪、黎文杰、黄敏、刘筱珉、张伟、徐龙奇、林华通、于欣、陈彦珺、宋超、宋素仟、黄浩、胡洋、熊鑫、梁国斌、刘泽坤、范司琪、朱均璐、杨斌、王冬东、谭斯梦、丁丽红、李沿、董鸣骏、崔凯、谢冤
结构工程师：师建伟、徐浩祥、杨洋、柴磊、付大伟、钟锡铭、金晓东、洪飞、朱芳、王莉娜、谢忠威、林逸风、吴强峰、徐泽恩、金明彦、王申昊、龚铭、严志威、钟奇、杨嘉伟、叶怀晨、赵凯龙、梁博俊
设备工程师：葛令科、杨福华、黄池钧、谷立芹、高利强、宋晓天、彭迎云、黄路、毛迪华、陈舟舟、曾杰、任庆军、郑铭、毛瑞琳、何骋宇、王倩雯、李程、高涵、王俊涛、刘丽芳、赵志铭、张雪祁、钱列东、寿广、叶金元、南旭、屠兴灿、陈文卉、魏山山、卢琦、孟娜、王雅迪
合作单位：
室内设计：梁志天设计
景观设计：SWA

Client: Hengli Group
Commission: Tender
Construction Brief: New Construction
Location: Suzhou, Jiangsu
Floor Area: 1,277,000m²
Design/Completion: 2020/-
Design Team:
Lead Architects: LING Jian, TIAN Yu, HAN Zhongqiang
Architecture: GUO Yin, ZHANG Tao, WANG Yong, KOU Guangjian, YU Xiang, JIANG Yin, YE Fan, ZHAO Lingyan, SHAO Xiaoqi, XIANG Pei, ZHENG Jue, JIA Yanqi, LI Wenjie, HUANG Min, LIU Xiaomin, ZHANG Wei, XU Longqi, LIN Huatong, YU Xin, CHEN Yanjun, SONG Chao, SONG Suqian, HUANG Hao, HU Yang, XIONG Xin, LIANG Guobin, LIU Zekun, FAN Siqi, ZHU Junlu, YANG Bin, WANG Dongdong, TAN Simeng, DING Lihong, LI Yan, DONG Mingjun, CUI Kai, XIE Mian
Structure: SHI Jianwei, XU Haoxiang, YANG Yang, CHAI Lei, FU Dawei, ZHONG Ximing, JIN Xiaodong, HONG Fei, ZHU Fang, WANG Lina, XIE Zhongwei, LIN Yifeng, WU Qiangfeng, XU Ze'en, JIN Mingyan, WANG Shenhao, GONG Ming, YAN Zhiwei, ZHONG Qi, YANG Jiawei, YE Huaichen, ZHAO Kailong, LIANG Bojun
MEP: GE Lingke, YANG Fuhua, HUANG Chijun, GU Liqin, GAO Liqiang, SONG Xiaotian, PENG Yingyun, HUANG Lu, MAO Dihua, CHEN Zhouzhou, ZENG Jie, REN Qingjun, ZHENG Ming, MAO Ruilin, HE Chengyu, WANG Qianwen, LI Cheng, GAO Han, WANG Juntao, LIU Lifang, ZHAO Zhiming, ZHANG Xueqi, QIAN Liedong, SHOU Guang, YE Jinyuan, NAN Xu, TU Xingcan, CHEN Wenhui, WEI Shanshan, LU Qi, MENG Na, WANG Yadi
Collaborators:
Interior: SLD
Landscape: SWA

杭州西动所上盖及周边区域综合开发
Hangzhou West High-speed Train Maintenance Base Superstructure

业主：浙江西铁交控房地产开发有限公司
项目来源：招投标
建设需求：新建
项目地点：浙江省杭州市
建设面积：410,000m²
设计 / 竣工：2022/—
设计团队：
项目负责：何兼、韩中强
建筑设计师：王勇、李凌、王冬东、丁丽红、高天、韩城、李沿、童杰、聂一蕾、孙寒啸、伍瞻颖、俞翔、张润东、徐子帆、陈志峰
结构工程师：徐浩祥、何亮、施冬、贾武鹏
设备工程师：任庆军、曾杰、叶金元
合作单位：
联合体单位：上海空间规划设计研究院有限公司

Client: Zhejiang West Railway Traffic Control Real Estate Development Co., Ltd.
Commission: Tender
Construction Brief: New Construction
Location: Hangzhou, Zhejiang
Floor Area: 410,000m²
Design/Completion: 2022/-
Design Team:
Lead Architects: HE Jian, HAN Zhongqiang
Architecture: WANG Yong, LI Ling, WANG Dongdong, DING Lihong, GAO Tian, HAN Cheng, LI Yan, TONG Jie, NIE Yilei, SUN Hanxiao, WU Zhanying, YU Xiang, ZHANG Rundong, XU Zifan, CHEN Zhifeng
Structure: XU Haoxiang, HE Liang, SHI Dong, JIA Wupeng
MEP: REN Qingjun, ZENG Jie, YE Jinyuan
Collaborator:
Consortium Member: Shanghai Spatial Planning and Design Institute

望江中心 TOD
Wangjiang Center TOD

业主：兴合集团
委托方式：直接委托
建设需求：新建＋改造
项目地点：浙江省杭州市
建筑面积：49,000m²
设计 / 竣工：2019/—
设计团队：
项目负责：张迅
建筑设计师：胡培、汪进、欧阳菊英、卢晓仪、郭绵沅津、童中拓、张增鑫、张文强
室内设计师：李扬、徐怀忱、赵怡洁、袁凯、沈冰洁、戴亮亮、林琳赟、胡文涛、黄牧舟
景观设计师：侯冬炜、张颖、郑涛、王羽
结构工程师：师建伟、严志威、贾武鹏
设备工程师：曾杰、黄琦、王兆星、王文胜、葛令科、吴文裕、卢琦、王超伟、寿广、张雪祁、赵睿、刘丽芳、钱列东

Client: Xinghe Group
Commission: Client Brief
Construction Brief: New Construction & Renovation
Location: Hangzhou, Zhejiang
Floor Area: 49,000m²
Design/Completion: 2019/-
Design Team:
Lead Architect: ZHANG Xun
Architecture: HU Pei, WANG Jin, OUYANG Juying, LU Xiaoyi, GUO Mianyuanjin, TONG Zhongtuo, ZHANG Zengxin, ZHANG Wenqiang
Interior: LI Yang, XU Huaichen, ZHAO Yijie, YUAN Kai, SHEN Bingjie, DAI Liangliang, LIN Linyun, HU Wentao, HUANG Muzhou
Landscape: HOU Dongwei, ZHANG Ying, ZHENG Tao, WANG Yu
Structure: SHI Jianwei, YAN Zhiwei, JIA Wupeng
MEP: ZENG Jie, HUANG Qi, WANG Zhaoxing, WANG Wensheng, GE Lingke, WU Wenyu, LU Qi, WANG Chaowei, SHOU Guang, ZHANG Xueqi, ZHAO Rui, LIU Lifang, QIAN Liedong

滨耀城
Colorful City

业主：滨江集团、兴耀房产
委托方式：直接委托
建设需求：新建
项目地点：浙江省杭州市
建筑面积：262,600m²
设计 / 竣工：2019/2023
设计团队：
项目负责：田钰
建筑设计师：寇广建、刘筱珉、吴一飞、杜立明、王俊、张辉、吴伟珍、宋素仟、吕博文、樊明朗、何儒迪、黄浩、陈勇、柴页新
结构工程师：黄伟志、庄新炉、赵晨、钟奇、叶怀晨、钟锡铭、陈聪、吴茂铭、胡天遥、杨嘉伟
设备工程师：杨福华、葛令科、曾杰、黄池钧、宋晓天、周益林、高利强、钱列东、郑铭、任庆军、李程、毛瑞琳、侯会芬、时云强、王客、寿广、叶金元、周佰明、屠兴灿、南旭、陈文卉、程磊、王超伟、卢琦
合作单位：
景观设计：棕榈设计
摄影：陈曦工作室

Client: Binjiang Real Estate, SHINION
Commission: Client Brief
Construction Brief: New Construction
Location: Hangzhou, Zhejiang
Floor Area: 262,600m²
Design/Completion: 2019/2023
Design Team:
Lead Architect: TIAN Yu
Architecture: KOU Guangjian, LIU Xiaomin, WU Yifei, DU Liming, WANG Jun, ZHANG Hui, WU Weizhen, SONG Suqian, LYU Bowen, FAN Minglang, HE Rudi, HUANG Hao, CHEN Yong, CHAI Yexin
Structure: HUANG Weizhi, ZHUANG Xinlu, ZHAO Chen, ZHONG Qi, YE Huaichen, ZHONG Ximing, CHEN Cong, WU Maoming, HU Tianyao, YANG Jiawei
MEP: YANG Fuhua, GE Lingke, ZENG Jie, HUANG Chijun, SONG Xiaotian, ZHOU Yilin, GAO Liqiang, QIAN Liedong, ZHENG Ming, REN Qingjun, LI Cheng, MAO Ruilin, HOU Huifen, SHI Yunqiang, WANG Ke, SHOU Guang, YE Jinyuan, ZHOU Weiming, TU Xingcan, NAN Xu, CHEN Wenhui, CHENG Lei, WANG Chaowei, LU Qi
Collaborators:
Landscape: Palm Design
Images: CHEN Xi Studio

天目里
OōEli

业主：慧展科技（杭州）有限公司
委托方式：直接委托
建设需求：新建
项目地点：浙江省杭州市
建筑面积：230,000m²
设计 / 竣工：2012/2020
设计团队：
项目负责：陆皓
建筑设计师：杜立明、卿州、刘琳、陈骏、王勇、胥小斌、裘敏、潘彬彬、陈舒、赵佳灵
室内设计师：林琳赟、诸双、戴亮亮、胡文涛、郭思聪、贾一帆、袁佳琛
结构工程师：胡凌华、包风、龚铭、陈森、杨洋、詹伟良、赵亮亮、周南、庞振钱、杜擎峰、王琦、敖国胜、周伟平
设备工程师：寿广、王超伟、程磊、陈文卉、庄少阳、王俊、曾菲、朱丹、何刚、叶金元、张雪祁、刘丽芳、侯会芬、张楠、李程、赵志铭、赵睿、赵莹、任丹华、钱列东、徐幸、刘佳莹、吕小斌、姚银杰、梅玉龙、王文胜、周益林、黄池钧、徐丽、陈舟舟、葛健、花勇刚、庞林华、葛令科、孙中南
合作单位：
总体规划：伦佐·皮亚诺建筑工作室
建筑设计：伦佐·皮亚诺建筑工作室
结构顾问：奥雅纳、goa 大象设计
机电顾问：奥雅纳、goa 大象设计
幕墙顾问：奥雅纳
植物顾问：Rana Creek
枯山水庭园顾问：枡野俊明 + 日本造园设计
水景顾问：JML 水景设计
景观深化设计：绿城爱境
灯光顾问：奥雅纳
声学顾问：SM&W 声美华
艺术顾问：弗朗切斯科·博纳米
标识顾问：SQPEG
摄影：Maxime LAURENT、goa 大象设计、天目里美术馆

Client: Huizhan Technology (Hangzhou) Co., Ltd.
Commission: Client Brief
Construction Brief: New Construction
Location: Hangzhou, Zhejiang
Floor Area: 230,000m²
Design/Completion: 2012/2020
Design Team:
Lead Architect: LU Hao
Architecture: DU Liming, QING Zhou, LIU Lin, CHEN Jun, WANG Yong, XU Xiaobin, QIU Min, PAN Binbin, CHEN Shu, ZHAO Jialing
Interior: LIN Linyun, ZHU Shuang, DAI Liangliang, HU Wentao, GUO Sicong, JIA Yifan, YUAN Jiachen
Structure: HU Linghua, BAO Feng, GONG Ming, CHEN Sen, YANG Yang, ZHAN Weiliang, ZHAO Liangliang, ZHOU Nan, PANG Zhenqian, DU Panfeng, WANG Qi, AO Guosheng, ZHOU Weiping
MEP: SHOU Guang, WANG Chaowei, CHENG Lei, CHEN Wenhui, ZHUANG Shaoyang, WANG Jun, ZENG Fei, ZHU Dan, HE Gang, YE Jinyuan, ZHANG Xueqi, LIU Lifang, HOU Huifen, ZHANG Yu, LI Cheng, ZHAO Zhiming, ZHAO Rui, ZHAO Ying, REN Danhua, QIAN Liedong, XU Xing, LIU Jiaying, LYU Xiaobin, YAO Yinjie, MEI Yulong, WANG Wensheng, ZHOU Yilin, HUANG Chijun, XU Li, CHEN Zhouzhou, GE Jian, HUA Yonggang, PANG Linhua, GE Lingke, SUN Zhongnan
Collaborators:
Planning: Renzo Piano Building Workshop
Architecture: Renzo Piano Building Workshop
Structure: Arup, GOA
M&E: Arup, GOA
Curtain Wall: Arup
Plant Consultant: Rana Creek Habitat Restoration
Japanese Garden: Shunmyo Masuno + Japan Landscape Consultants
Waterscape: JML Water Feature Design
Landscape Design Development: Greentown Akin
Lighting: Arup
Acoustic: Shen Milsom & Wilke
Art Consultant: Francesco Bonami
Signage: Square Peg Design
Images: Maxime LAURENT, GOA, By Art Matters

浙商银行总部
China Zheshang Bank Headquarters

业主：浙商银行
委托方式：招投标
建设需求：新建
项目地点：浙江省杭州市
建筑面积：310,000m²
设计 / 竣工：2018/—
设计团队：
项目负责：陆皓、陈娴
建筑设计师：凌建、陈斌鑫、徐琦、张琪琳、陈吉、蒋嘉菲、杜立明、庄雪松、刘安、陈致浩、周星宇、杜皓月、刘宇、张智运、吴伟珍、刘波、郝博文、胡夫遥、秦阗怡、胡阳春、吴敬波、钱杰
景观设计师：侯冬炜、姜欢、朱靜、郑涛
结构工程师：胡凌华、徐浩祥、何亮、龚铭、杨洋、宋子文、吴涅峰、陈聪、付大伟
设备工程师：寿广、南旭、程磊、陈文卉、魏山山、卢琦、周伟明、叶金元、任庆军、刘源、王俊涛、王信雯、何骋宇、张雪祁、钱列东、葛令科、谷立芹、周益林、王兆星、施昱展、杨福华、曾杰、徐幸、姚银杰
BIM 工程师：朱理一、时云强、许喆、王凯烽、任庆阳、王本利、高涵、盛梦云、屠兴灿
合作单位：
结构顾问：华东建筑设计研究院
机电顾问：SM 思迈
幕墙顾问：奥雅纳
摄影：goa 大象设计

Client: China Zheshang Bank
Commission: Tender
Construction Brief: New Construction
Location: Hangzhou, Zhejiang
Floor Area: 310,000m²
Design/Completion: 2018/-
Design Team:
Lead Architect: LU Hao, CHEN Xian
Architecture: Lingjian, CHEN Binxin, XU Qi, ZHANG Qilin, CHEN Ji, JIANG Jiafei, DU Liming, ZHUANG Xuesong, LIU An, CHEN Zhihao, ZHOU Xingyu, DU Haoyue, LIU Yu, ZHANG Zhiyun, WU Weizhen, LIU Bo, HAO Bowen, HU Tianyao, QIN Tianyi, HU Yangchun, WU Jingbo, QIAN Jie
Landscape: HOU Dongwei, JIANG Huan, ZHU Jing, ZHENG Tao
Structure: HU Linghua, XU Haoxiang, HE Liang, GONG Ming, YANG Yang, SONG Ziwen, WU Qiangfeng, CHEN Cong, FU Dawei
MEP: SHOU Guang, NAN Xu, CHENG Lei, CHEN Wenhui, WEI Shanshan, LU Qi, ZHOU Weiming, YE Jinyuan, REN Qingjun, LIU Yuan, WANG Juntao, WANG Qianwen, HE Chengyu, ZHANG Xueqi, QIAN Liedong, GE Lingke, GU Liqin, ZHOU Yilin, WANG Zhaoxing, SHI Yuzhan, YANG Fuhua, ZENG Jie, XU Xing, YAO Yinjie
BIM: ZHU Liyi, SHI Yunqiang, XU Zhe, WANG Kaifeng, REN Qingyang, WANG Benli, GAO Han, SHENG Mengyun, TU Xingcan
Collaborators:
Structure: ECADI
M&E: Squire Mech
Curtain Wall: Arup
Images: GOA

宇视科技总部
Uniview Headquarters

业主：宇视科技
委托方式：直接委托
建设需求：新建
项目地点：浙江省杭州市
建筑面积：136,700m²
设计 / 竣工：2017/2023
设计团队：
项目负责：陈斌鑫
建筑设计师：郑文康、谷裕、徐艳、刘安、季湘志、李潇乐、蒋经军、尹建博、石逸雄、徐雅甜
结构工程师：师建伟、陈森、朱芳、陈聪、龚铭、敖国胜、吴茂铭、徐泽恩、赵圣民
设备工程师：黄路、沈海勤、施昱展、葛令科、曾杰、侯会芬、任庆军、王客、何骋宇、张雪祁、寿广、南旭、程磊、魏山山、王超伟、叶金元、徐幸、姚银杰
合作单位：
景观设计：绿风生态旅游规划设计研究院
摄影：RudyKu

Client: Uniview
Commission: Client Brief
Construction Brief: New Construction
Location: Hangzhou, Zhejiang
Floor Area: 136,700m²
Design/Completion: 2017/2023
Design Team:
Lead Architect: CHEN Binxin
Architecture: ZHENG Wenkang, GU Yu, XU Yan, LIU An, JI Xiangzhi, LI Xiaole, JIANG Jingjun, YIN Jianbo, SHI Yixiong, XU Yatian
Structure: SHI Jianwei, CHEN Sen, ZHU Fang, CHEN Cong, GONG Ming, AO Guosheng, WU Maoming, XU Ze'en, ZHAO Shengmin
MEP: HUANG Lu, SHEN Haiqin, SHI Yuzhan, GE Lingke, ZENG Jie, HOU Huifen, REN Qingjun, WANG Ke, HE Chengyu, ZHANG Xueqi, SHOU Guang, NAN Xu, CHENG Lei, WEI Shanshan, WANG Chaowei, YE Jinyuan, XU Xing, YAO Yinjie
Collaborators:
Landscape: Green Wind Eco-tourism Planning and Design Institute
Images: RudyKu

西子智慧产业园
Hangzhou Xizi Wisdom Industrial Park

业主：西子联合
委托方式：直接委托
建设需求：新建 + 改造
项目地点：浙江省杭州市
建筑面积：310,000m²
设计 / 竣工：2018/2023
设计团队：
项目负责：凌建、陈吉
建筑设计师：李洪、钱炎辉、刘大可、朱尧、陈韫、蒋林峰、管亮、宣蕊、郑亚军、茅吉祥
结构工程师：黄伟志、詹伟良、庄新炉、汪卓红、王立才、周孝志、冯小生、董超、宋鹊、崔碧琪
设备工程师：黄池钧、周益林、杨福华、王晓卉、曾杰、葛令科、宋晓天、梅玉龙、黄文杰、毛迪华、魏民、王俊涛、吴金祥、钱列东、赵莹、张宇明、李江龙、郑铭、李程、毛瑞琳、刘丽芳、黄钦鹏、寿广、吴云裕、叶金元、王雍迪、朱应钦
合作单位：
景观设计：同创工程设计有限公司
摄影：goa 大象设计、泠城摄影工作室

Client: Xizi UHC
Commission: Client Brief
Construction Brief: New Construction & Renovation
Location: Hangzhou, Zhejiang
Floor Area: 310,000m²
Design/Completion: 2018/2023
Design Team:
Lead Architect: LING Jian, CHEN Ji
Architecture: LI Hong, QIAN Yanhui, LIU Dake, ZHU Yao, CHEN Yun, JIANG Linfeng, GUAN Liang, XUAN Rui, ZHENG Yajun, MAO Jixiang
Structure: HUANG Weizhi, ZHAN Weiliang, ZHUANG Xilu, WANG Zhuohong, WANG Licai, ZHOU Xiaozhi, FENG Xiaosheng, DONG Chao, CUI Biqi
MEP: HUANG Chijun, ZHOU Yilin, YANG Fuhua, WANG Wensheng, WANG Xiaohui, ZENG Jie, GE Lingke, SONG Xiaotian, MEI Yulong, HUANG Wenjie, MAO Dihua, WEI Min, WANG Juntao, WU Jinxiang, QIAN Liedong, ZHAO Ying, ZHANG Yuming, LI Jianglong, ZHENG Ming, LI Chen, MAO Ruilin, LIU Lifang, HUANG Qinpeng, SHOU Guang, WU Wenyu, YE Jinyuan, WANG Yadi, ZHU Yingqin
Collaborators:
Landscape: Tongchuang Engineering Design Co., Ltd.
Images: GOA, SHIROMIO Studio

杭州东站花园国际
Hangzhou East Railway Station Garden International

业主：杭州花园股份经济合作社
委托方式：直接委托
建设需求：新建
项目地点：浙江省杭州市
建筑面积：138,500m²
设计 / 竣工：2017/2022
设计团队：
项目负责：凌建、张琪琳
建筑设计师：陈吉、吴敬波、吴若晨、赵书艺、冯玉清
结构工程师：师建伟
设备工程师：葛令科、刘源、王超伟
合作单位：
施工图设计：平安建设
摄影：RudyKu

Client: Hangzhou Huayuan Joint-equity Economic Cooperative
Commission: Client Brief
Construction Brief: New Construction
Location: Hangzhou, Zhejiang
Floor Area: 138,500m²
Design/Completion: 2017/2022
Design Team:
Lead Architects: LING Jian, ZHANG Qilin
Architecture: CHEN Ji, WU Jingbo, WU Ruochen, ZHAO Shuyi, FENG Yuqing
Structure: SHI Jianwei
MEP: GE Lingke, LIU Yuan, WANG Chaowei
Collaborators:
Construction Drawing: Ping An Construction
Images: RudyKu

海口五源河创新产业中心
Haikou Wuyuanhe Intelligent Creative Collective

业主：海南锦绣实业有限公司
委托方式：直接委托
建设需求：新建
项目地点：海南省海口市
建筑面积：415,000m²
设计 / 竣工：2021/—
设计团队：
项目负责：凌建、陈吉
建筑设计师：周星宇、陈禹男、王浩名、谭斯梦、施一豪、刘锦洲
结构工程师：徐浩祥、杨洋、朱芳
设备工程师：王超伟、孟娜、南旭、寿广、杨福华、黄路、毛迪华、钱列东、刘源、刘丽芳
合作单位：
室内设计：易和设计
景观设计：ACA 麦垦景观
施工图设计：柏森设计

Client: Hainan Jinxiu Industrial Co., Ltd.
Commission: Client Brief
Construction Brief: New Construction
Location: Haikou, Hainan
Floor Area: 415,000m²
Design/Completion: 2021/-
Design Team:
Lead Architects: LING Jian, CHEN Ji
Architecture: ZHOU Xingyu, CHEN Yunan, TAN Simeng, SHI Yihao, LIU Jinzhou
Structure: XU Haoxiang, YANG Yang, ZHU Fang
MEP: WANG Chaowei, MENG Na, NAN Xu, SHOU Guang, YANG Fuhua, HUANG Lu, MAO Dihua, QIAN Liedong, LIU Yuan, LIU Lifang
Collaborators:
Interior: EH Design
Landscape: Aicon Landscape
Construction Drawing: Person Design

江南布衣仓储园区
JNBY Warehousing Logistics Park

业主：江南布衣
委托方式：直接委托
建设需求：新建
项目地点：浙江省杭州市
建筑面积：91,000m²
设计 / 竣工：2016/2019
设计团队：
项目负责：陆皓、王彦
建筑设计师：夏斐、杜立明、胥小斌、杜宝山、李夫龙、殷长伟
结构工程师：师建伟、龚铭、张建辉、柴磊、叶小明、林逸风
设备工程师：王文胜、高利强、陈舟舟、梅玉龙、寿广、叶金元、王超伟、钱列东、魏民、张雪祁、侯会芬、徐幸、姚银杰
合作单位：
摄影：吕恒中

Client: JNBY
Commission: Client Brief
Construction Brief: New Construction
Location: Hangzhou, Zhejiang
Floor Area: 91,000m²
Design/Completion: 2016/2019
Design Team:
Lead Architects: LU Hao, WANG Yan
Architecture: XIA Fei, DU Liming, XU Xiaobin, DU Baoshan, LI Fulong, YIN Changwei
Structure: SHI Jianwei, GONG Ming, ZHANG Jianhui, CHAI Lei, YE Xiaoming, LIN Yifeng
MEP: WANG Wensheng, GAO Liqiang, CHEN Zhouzhou, MEI Yulong, SHOU Guang, YE Jinyuan, WANG Chaowei, QIAN Liedong, WEI Min, ZHANG Xueqi, HOU Huifen, XU Xing, YAO Yinjie
Collaborators:
Images: LYU Hengzhong

石家庄中央商务区办公楼
Shijiazhuang CBD Office Towers

业主：石家庄市中央商务区开发有限公司
委托方式：招投标
建设需求：新建
项目地点：河北省石家庄市
建筑面积：136,800m²
设计 / 竣工：2019/—
设计团队：
项目负责：陆皓、韩中强
建筑设计师：胡洋、刘泽坤、范司琪、孙寒啸、罗胜之、孙煜
结构工程师：师建伟、杨洋
设备工程师：杨福华、吴文裕、魏民、黄池钧、刘丽芳
合作单位：
室内设计：CCD 郑中设计事务所
景观设计：澳派景观设计工作室
施工图设计：北京市建筑设计研究院
摄影：RudyKu

Client: Shijiazhuang Central Business District Development Co., Ltd.
Commission: Tender
Construction Brief: New Construction
Location: Shijiazhuang, Hebei
Floor Area: 136,800m²
Design/Completion: 2019/-
Design Team:
Lead Architects: LU Hao, HAN Zhongqiang
Architecture: HU Yang, LIU Zekun, FAN Siqi, SUN Hanxiao, LUO Shengzhi, SUN Yu
Structure: SHI Jianwei, YANG Yang
MEP: YANG Fuhua, WU Wenyu, WEI Min, HUANG Chijun, LIU Lifang
Collaborators:
Interior: Cheng Chung Design
Landscape: ASPECT Studios
Construction Drawing: BIAD
Images: RudyKu

台州数字科技园
Taizhou Digital Technology Park

业主：台州市城投科技发展有限公司
委托方式：招投标
建设需求：新建
项目地点：浙江省台州市
建筑面积：231,450m²
设计 / 竣工：2022/—
设计团队：
项目负责：何兼、韩中强
建筑设计师：李宏捷、崔凯、刘泽坤、蒋洒洒、徐子帆、莫利强、江雨蓉、戴超、吴婷婷、向陪、余偲聪、邵笑琦
室内设计：李扬、袁凯
景观设计师：侯冬炜、张颖、邢益楠、徐怀忱
结构工程师：师建伟、杨洋、龚铭、吴强峰、钟奇、陈聪、王申昊、俞洪、汪卓红、施冬、谢忠威、王本利
设备工程师：寿广、程磊、陈文卉、孟娜、叶金元、葛令科、高利强、谷立芹、徐丽、宋晓天、周益林、杨福华、任庆军、黄钦鹏、侯会芬、时云强、张雪祁、钱列东、徐幸、姚银杰
BIM 工程师：时云强、高涵、卢琦、盛梦云、王兆星、徐丽、姚远

Client: Taizhou Chengtou Technology Development Co., Ltd.
Commission: Tender
Construction Brief: New Construction
Location: Taizhou, Zhejiang
Floor Area: 231,450m²
Design/Completion: 2022/-
Design Team:
Lead Architects: HE Jian, HAN Zhongqiang
Architecture: LI Hongjie, CUI Kai, LIU Zekun, JIANG Sasa, XU Zifan, MO Liqiang, JIANG Yurong, DAI Chao, WU Tingting, XIANG Pei, YU Sicong, SHAO Xiaoqi
Interior: Li Yang, Yuan Kai
Landscape: Hou Dongwei, Zhang Ying, Xing Yinan, Xu Huaichen
Structure: SHI Jianwei, YANG Yang, GONG Ming, WU Qiangfeng, ZHONG Qi, CHEN Cong, WANG Shenhao, YU Hong, WANG Zhuohong, SHI Dong, XIE Zhongwei, WANG Benli
MEP: SHOU Guang, CHENG Lei, CHEN Wenhui, MENG Na, YE Jinyuan, GE Lingke, GAO Liqiang, GU Liqin, XU Li, SONG Xiaotian, ZHOU Yilin, YANG Fuhua, REN Qingjun, HUANG Qinpeng, HOU Huifen, SHI Yunqiang, ZHANG Xueqi, QIAN Liedong, XU Xing, YAO Yinjie
BIM: SHI Yunqiang, GAO Han, LU Qi, SHENG Mengyun, WANG Zhaoxing, XU Li, YAO Yuan

宁波智造港芯创园
Ningbo Intelligent Manufacturing Port

业主：绿城中国
项目来源：直接委托
建设需求：新建
项目地点：浙江省宁波市
建筑面积：310,000m²
设计 / 竣工：2020/—
设计团队：
项目负责：凌建、蒋嘉菲
建筑设计师：邹洁琳、张弦、陈吉、李洪、刘大可、兰小明、钱焱辉、林可峰、周星宇、林宗雄、凌利平、黄鹤、王浩明、李平原
结构工程师：柴磊
设备工程师：杨福华、吴金祥、南旭
合作单位：
景观设计：浙江蓝颂园林景观设计集团有限公司
施工图设计：宁波市房屋建筑设计研究院有限公司
摄影：一乘摄影

Client: Greentown China
Commission: Client Brief
Construction Brief: New Construction
Location: Ningbo, Zhejiang
Floor Area: 310,000m²
Design/Completion: 2020/-
Design Team:
Lead Architects: LING Jian, JIANG Jiafei
Architecture: ZOU Jielin, ZHANG Xian, CHEN Ji, LI Hong, LIU Dake, LAN Xiaoming, QIAN Yanhui, LIN Kefeng, ZHOU Xingyu, LIN Zongxiong, LING Liping, HUANG He, WANG Haoming, LI Pingyuan
Structure: CHAI Lei
MEP: YANG Fuhua, WU Jinxiang, NAN Xu
Collaborators:
Landscape: GTS Lansong Design Group
Construction Drawing: Ningbo Building Design Institute Co., Ltd.
Images: Yi Cheng Studio

祥符桥传统风貌街区
Xiangfu Bridge Historic District Renewal

业主：杭州市拱墅区城市建设发展中心
委托方式：招投标
建设需求：新建 + 改造
项目地点：浙江省杭州市
建筑面积：66,100m²
设计 / 竣工：2019/2023
设计团队：
项目负责：张晓晓、陈斌鑫、张迅、王彦
建筑设计师：王忠杰、郑文康、杨必龙、于军贵、季湘志、刘安、祝志根、徐艳、刘丹荔、方言智、胡一帆、薛乙浩、李阳、贾高松、叶茂华、梅富鹏、尹建博、石逸雄、方婷、汪进、江昊、刘洪海、卢晓仪、宋超、黄敏、吴超楠、郑静云、彭书勉、窦志国、孙钰琪
室内设计师：高雁静智、任丽娜、毛瑜玫、袁佳琛
景观设计师：侯冬炜、程欣、李亚萍、李紫薇、袁美云、杨怡然、郑涛、黄佳英、薛天炜、王羽、卓百会、张颖、邢益楠、赵犟珅、姜爽
结构工程师：包风、杨泽、王琦、汪卓红、谢忠威、任庆阳、王本利、陈聪、施冬、宋子文、吴强峰、何亮、陈森、莫一奇、高俊、金明彦
设备工程师：程磊、魏山山、寿广、吴文裕、刘源、何骋宇、张雪祁、钱列东、黄路、沈海勤、施昱展、葛令科、杨福华、曾杰、吕小斌、姚银杰
BIM 工程师：卢沈杰、张振华、许喆、徐天钧
EPC 管理：沈旭凯、吕婷婷、高俊、童玥、魏民、吴永芳、张科、孙昊、王周峰、王创达、王晓晴、马长春、冯丽琴、卢琳、孙杭金
合作设计：
景观设计：TOPO
摄影：奥观建筑视觉

Client: Hangzhou Gongshu District Urban Construction and Development Center
Commission: Tender
Construction Brief: New Construction & Renovation
Location: Hangzhou, Zhejiang
Floor Area: 66,100m²
Design/Completion: 2019/2023
Design Team:
Lead Architects: Shawn CHEUNG, CHEN Binxin, ZHANG Xun, WANG Yan,
Architecture: WANG Zhongjie, ZHENG Wenkang, YANG Bilong, YU Jungui, JI Xiangzhi, Liu An, ZHU Zhigen, XU Yan, LIU Danli, FANG Yanzhi, HU Yifan, XUE Yihao, LI Yang, JIA Gaosong, YE Maohua, MEI Fupeng, YIN Jianbo, SHI Yixiong, FANG Ting, WANG Jing, JIANG Hao, LIU Hongmei, LU Xiaoyi, SONG Chao, HUANG Ming, WU Chaonan, ZHENG Jingyun, PENG Shumian, DOU Zhiguo, SUN Yuqi
Interior: Gao Yajingzhi, REN Lina, MAO Yumei, YUAN Jiachen
Landscape: HOU Dongwei, CHENG Xin, LI Yaping, LI Ziwei, YUAN Meiyun, YANG Yiran, ZHENG Tao, HUANG Jiayin, XUE Tianwei, WANG Yu, ZHUO Baihui, ZHANG Ying, XINGYinan, ZHAO Jianshen, JIANG Huan
Structure: BAO Feng, YANG Yang, WANG Qi, WANG Zhuohong, XIE Zhongwei, REN Qingyang, WANG Benli, CHEN Cong, SHI Dong, SONG Ziwen, WU Qiangfeng, HE Liang, CHEN Sen, MO Yiqi, GAO Jun, JIN Mingyan
MEP: CHENG Lei, WEI Shanshan, SHOU Guang, WU Wenyu, LIU Yuan, HE Chengyu, ZHANG Xueqi, QIAN Liedong, HUANG Lu, SHEN Haiqin, SHI Yuzhan, GE Lingke, YANG Fuhua, ZENG Jie, XU Xing, LYU Xiaobin, YAO Yinjie
BIM: LU Shenjie, ZHANG Zhenhua, XU Zhe, XU Tianjun
EPC: SHEN Xukai, LYU Tingting, GAO Jun, TONG Yue, WEI Min, WU Yongfang, ZHANG Ke, SUN Hao, WANG Zhoufeng, WANG Chuangda, WANG Xiaoqing, MA Changchun, FENG Liqin, LU Lin, SUN Hangjin
Collaborators:
Landscape: TOPO
Images: AOGVISION

上海北外滩32街坊更新
Shanghai North Bund Neighborhood 32 Renewal

业主：新湖地产
项目来源：委托
建设需求：新建
项目地点：上海市虹口区
建筑面积：112,500m²
设计 / 竣工：2020/
设计团队：
项目负责：陆皓、梁卓敏
建筑设计师：李政、潘奕铭、吴金英、杜皓月、宓芬、杨玉琳、李梓琳、吴世旭、曹慧、吴鹏
结构工程师：徐浩祥
设备工程师：任庆军、何骞、徐幸、葛令科、叶金元
合作单位：
室内设计顾问：李裕棠
室内设计：三圣建筑
景观设计：PLA 景观设计
施工图设计：上海中房建筑设计有限公司
摄影：陈曦工作室

Client: XINHU Real Estate
Commission: Client Brief
Construction Brief: New Construction
Location: Hongkou, Shanghai
Floor Area: 112,500m²
Design/Completion: 2020/
Design Team:
Lead Architects: LU Hao, LIANG Zhuomin
Architecture: LI Zheng, PAN Yiming, WU Jinying, DU Haoyue, MI Fen, YANG Yulin, LI Zilin, WU Shixu, CAO Hui, WU Peng
Structure: XU Haoxiang
MEP: REN Qingjun, HE Qian, XU Xing, GE Lingke, YE Jinyuan
Collaborators:
Interior Consultant: TonyChi
Interior: Design333
Landscape: PLA
Construction Drawing: Shanghai ZF Architectural Design Co., LTD
Images: CHEN Xi Studio

弘安里
Hong'anli

业主：绿城中国
委托方式：直接委托
建设需求：新建 + 改造
项目地点：上海市虹口区
建筑面积：95,200m²
设计 / 竣工：2021/—
设计团队：
项目负责：陆皓、梁卓敏、刘波
建筑设计师：秦阗怡、余喆偲、蒋骁闻、鲍华英、祁雅菁、吴世旭、李梓林、徐正、边雪松、宓芬、林际远、李富贵、孙佳豪、王冉星、施忆
结构工程师：徐浩祥、柴磊
设备工程师：葛令科、任庆军、叶金元
合作单位：
室内设计：CCD 郑中设计事务所、无间设计、杭州大诠建筑装饰设计有限公司
景观设计：DLC 地茂景观设计、GTS 蓝颂设计集团
施工图设计：天华
摄影：陈曦工作室

Client: Greentown China
Commission: Client Brief
Construction Brief: New Construction & Renovation
Location: Hongkou, Shanghai
Floor Area: 95,200m²
Design/Completion: 2021/-
Design Team:
Lead Architects: LU Hao, LIANG Zhuomin, LIU Bo
Architecture: QIN Tianyi, YU Zhesi, JIANG Xiaowen, BAO Huaying, QI Yajing, WU Shixu, LI Zilin, XU Zheng, ZHUANG Xuesong, MI Fen, LIN Jiyuan, LI Fugui, SUN Jiahao, WANG Ranxing, SHI Yi
Structure: XU Haoxiang, CHAI Lei
MEP: GE Lingke, REN Qingjun, YE Jinyuan
Collaborators:
Interior: Cheng Chung Design, W.DESIGN, Hangzhou Daquan Architectural Decoration Design Co., Ltd.
Landscape: Design Land Collaborative, GTS Lansong Design Group
Construction Drawing: TIANHUA
Images: CHEN Xi Studio

中海顺昌玖里
China Overseas Arbour

业主：中海地产
项目来源：直接委托
建设需求：新建 + 改造
项目地点：上海市黄浦区
建筑面积：280,000m²
设计 / 竣工：2020/—
设计团队：
项目负责：袁源、王彦、李震
建筑设计师：徐勤力、胡祖龙、俞龙捷、张建宇、曹力扬、杨新宇、吴圣晗、蔡瑞、张漠、郑静云、徐震歆、熊志邦、陆薇薇、刘杜娟、施鹏骅、刘子文、殷长伟、苏子茗
结构工程师：徐浩祥、严志威、贾武鹏
设备工程师：张雪祁、曾杰、程磊
合作单位：
室内设计：李玮珉建筑师事务所 + 上海越界
景观设计：太璞建筑环境设计咨询（上海）有限公司
施工图设计：上海中房建筑设计有限公司
历史建筑保护顾问：华东建筑设计研究院有限公司历史建筑保护设计院、上海章明建筑设计事务所

Client: CHINA OVERSEAS
Commission: Client Brief
Construction Brief: New Construction & Renovation
Location: Huangpu, Shanghai
Floor Area: 280,000m²
Design/Completion: 2020/-
Design Team:
Lead Architects: Yuan Yuan, WANG Yan, LI Zhen
Architecture: XU Qinli, HU Zulong, YU Longjie, ZHANG Jianyu, CAO Liyang, YANG Xinyu, WU Shenghan, CAI Rui, Zhang Hao, ZHENG Jingyun, XU Chenxin, XIONG Zhidan, LU Weiwei, LIU Dujuan, SHI Penghua, LIU Ziwen, YIN Changwei, SU Ziming
Structure: XU Haoxiang, YAN Zhiwei, JIA Wupeng
MEP: ZHANG Xueqi, Zengjie, CHENG Lei
Collaborators:
Interior: LWMA × Shanghai Yue Jie
Landscape: TOPO
Construction Drawing: Shanghai ZF Architectural Design Co., LTD
Historic Preservation Consultant: ECADI, Shanghai Zhang Ming Architectural Design Co., LTD

里直街
Lizhi Street

业主：浙江省上虞曹娥江旅游度假区管理委员会
委托方式：直接委托
建设需求：改造
项目地点：浙江省绍兴市
建筑面积：4,300m²
设计 / 竣工：2019/2022
设计团队：
项目负责：陈斌鑫
建筑设计师：郑文康、袁波、刘安、祝志根、徐艳、王磬然、季湘志、余偲聪、尹建博、吴岳啸、刘广荣、卢耿聪、刘涵、石逸雄
结构工程师：徐浩祥、陈洪刚、钟锡铭、赵圣民、于悦
设备工程师：叶金元、王超伟、周伟明、卢琦、毛迪华、周益林、黄路、葛令科、张雪祁、侯会芬、刘源
合作单位：
景观设计：杭州蓝颂园林景观设计有限公司
摄影：goa 大象设计

Client: Cao'e River Tourist Resorts Administrative Committee of Shangyu, Zhejiang Province
Commission: Client Brief
Construction Brief: Renovation
Location: Shaoxing, Zhejiang
Floor Area: 4,300m²
Design/Completion: 2019/2022
Design Team:
Lead Architect: CHEN Binxin
Architecture: ZHENG Wenkang, YUAN Bo, LIU An, ZHU Zhigen, XU Yan, WANG Muran, JI Xiangzhi, YU Sicong, YIN Jianbo, WU Yuexiao, LIU Guangrong, LU Gengcong, LIU Han, SHI Yixiong
Structure: XU Haoxiang, CHEN Honggang, ZHONG Ximing, ZHAO Shengmin, YU Yue
MEP: YE Jinyuan, WANG Chaowei, ZHOU Weiming, LU Qi, MAO Dihua, ZHOU Yilin, HUANG Lu, GE Lingke, ZHANG Xueqi, HOU Huifen, LIU Yuan
Collaborators:
Landscape Design: GTS Lansong Design Group
Images: GOA

绿城湖境云庐
Greentown Hangzhou Oriental Villa

业主：绿城中国
委托方式：直接委托
建设需求：新建
项目地点：浙江省杭州市
建筑面积：146,100m²
设计 / 竣工：2018/2021
设计团队：
项目负责：何兼、袁源
建筑设计师：董慧、王宇、叶帆、白树全、夏杰、俞双娟、翁伯琛、程青依
结构工程师：徐浩祥、何亮、林逸风、董忆夏、宋子文、贾武鹏、施冬、于悦
设备工程师：周伟明、程磊、王俊、叶金元、钱列东、张雪祁、刘丽芳、李程、赵志铭、梅玉龙、王文胜、陈舟舟、彭迎云
合作单位：
室内设计：HWCD
景观设计：CLD 会筑景观
摄影：泠城摄影工作室、绿城中国

Client: Greentown China
Commission: Client Brief
Construction Brief: New Construction
Location: Hangzhou, Zhejiang
Floor Area: 146,100 m²
Design/Completion: 2018/2021
Design Team:
Lead Architects: HE Jian, YUAN Yuan
Architecture: DONG Hui, WANG Yu, YE Fan, BAI Shuquan, XIA Jie, YU Shuangjuan, WENG Bozhang, CHENG Qingyi
Structure: XU Haoxiang, HE Liang, LIN Yifeng, DONG Yixia, SONG Ziwen, JIA Wupeng, SHI Dong, YU Yue
MEP: ZHOU Weiming, CHENG Lei, WANG Jun, YE Jinyuan, QIAN Liedong, ZHANG Xueqi, LIU Lifang, LI Cheng, ZHAO Zhiming, MEI Yulong, WANG Wensheng, CHEN Zhouzhou, PENG Yingyun
Collaborators:
Interior: Harmony World Consultant & Design
Landscape: Collective Landscape Design
Images: SHIROMIO Studio, Greentown China

绿地海珀外滩
Greenland Hysun Bund

业主：绿地控股
委托方式：直接委托
建设需求：新建
项目地点：上海市黄浦区
建筑面积：208,000m²
设计 / 竣工：2015/—
设计团队：
项目负责：何兼、袁源、韩中强
建筑设计师：窦志国、方婷、陈俊、李雯雯、蒋寅、汪海莹、李瑾、邵笑琦、李素萍、白树全
结构工程师：徐浩祥、金明彦、严志威、柴磊、何亮、丁浩、俞晶晶、陈优优、宋子文、冯小生
设备工程师：葛令科、王兆星、周益林、谷立芹、曾杰、梅玉龙、魏民、何骞、刘丽芳、王倩雯、钟跃进、李文璧、王俊涛、郑铭、钱列东、叶金元、朱丹、王雅迪、邓雅琼、寿广、徐奎、刘佳莹
合作单位：
室内设计：维几设计、集艾设计、HBA、上海成高建筑装饰设计有限公司
景观设计：纳千景观
摄影：田方方

Client: Greenland Holdings
Commission: Client Brief
Construction Brief: New Construction
Location: Huangpu, Shanghai
Floor Area: 208,000m²
Design/Completion: 2015/-
Design Team:
Lead Architects: HE Jian, YUAN Yuan, HAN Zhongqiang
Architecture: DOU Zhiguo, FANG Ting, CHEN Jun, LI Wenwen, JIANG Yin, WANG Haiying, LI Jin, SHAO Xiaoqi, LI Suping, BAI Shuquan
Structure: XU Haoxiang, JIN Mingyan, YAN Zhiwei, CHAI Lei, HE Liang, DING Hao, YU Jingjing, CHEN Youyou, SONG Ziwen, FENG Xiaosheng
MEP: GE Lingke, WANG Zhaoxing, ZHOU Yilin, GU Liqin, ZENG Jie, MEI Yulong, WEI Min, HE Qian, LIU Lifang, WANG Qianwen, ZHONG Yuejin, LI Wenbi, WANG Juntao, ZHENG Ming, QIAN Liedong, YE Jinyuan, ZHU Dan, WANG Yadi, DENG Yaqiong, SHOU Guang, XU Xing, LIU Jiaying
Collaborators:
Interior: WJID, G-Art Design, Hirsch Bedner Associates, Shanghai Chenggao Construction Decoration Design Co., Ltd.
Landscape: LGWS Design
Images: Studio FF

仁恒海上源
Yanlord Arcadia

业主：仁恒置地
委托方式：直接委托
建设需求：新建
项目地点：上海市杨浦区
建筑面积：292,400m²
设计 / 竣工：2018/—
设计团队：
项目负责：陆皓、张琪琳
建筑设计师：吴宁宁、吴敬波、张蓓艳、张淏、辛博文、陆薇薇、关伟超、王新业、邱立文、赵书艺
合作单位：
室内设计：李玮珉建筑师事务所、MDO 木君建筑设计、无间设计、维几设计
景观设计：TROP、魏玛景观
施工图设计：天华
摄影：清筑影像

Client: Yanlord Land
Commission: Client Brief
Construction Brief: New Construction
Location: Yangpu, Shanghai
Floor Area: 292,400m²
Design/Completion: 2018/-
Design Team:
Lead Architects: LU Hao, ZHANG Qilin
Architecture: WU Ningning, WU Jingbo, ZHANG Beiyan, ZHANG Hao, XIN Bowen, LU Weiwei, GUAN Weichao, WANG Xinye, SHAO Liwen, ZHAO Shuyi
Collaborators:
Interior: LWM Architects, More Design Office, W.DESIGN, WJID
Landscape: TROP: terrains + open space, WEIMAR
Construction Drawing: TIANHUA
Images: CreatAR

融创长乐雅颂
Sunac Changle Yasong

业主：融创中国
委托方式：直接委托
建设需求：新建
项目地点：重庆市巴南区
建筑面积：108,000m²
设计 / 竣工：2019/—
设计团队：
项目负责：何峻、周羿
建筑设计师：朱欣伟、秦小元、陈融、卞弢、魏炜
结构工程师：俞洪
设备工程师：葛令科、吴文裕
合作单位：
室内设计：维几设计
景观设计：种地设计
施工图设计：AAD 长厦安基
摄影：goa 大象设计

Client: Sunac
Commission: Client Brief
Construction Brief: New Construction
Location: Banan, Chongqing
Floor Area: 108,000m²
Design/Completion: 2019/-
Design Team:
Lead Architects: HE Jun, ZHOU Yi
Architecture: ZHU Xinwei, QIN Xiaoyuan, CHEN Rong, BIAN Tao, WEI Wei
Structure: YU Hong
MEP: GE Lingke, WU Wenyu
Collaborators:
Interior: WJID
Landscape: ZhongDi Design
Construction Drawing: AAD
Images: GOA

华润亚奥城
CR Land the Century City

业主：华润置地
委托方式：直接委托
建设需求：新建
项目地点：浙江省杭州市
建筑面积：486,650m²
设计 / 竣工：2018/2023
设计团队：
项目负责：张晓晓、韩中强
建筑设计师：王忠杰、胡洋、胡一帆、薛乙浩、李阳、刘卓星、贾高松、刘泽坤、范司琪、孙836晓啸、玉镇珲、蒋鹏飞、郭吟
结构工程师：黄伟志、庄新炉、严志威、董胜民
设备工程师：葛令科、谷立芹、寿广、南旭、刘源
合作单位：
总体规划：KPF
景观设计：JCFO、LANDAU 朗道国际设计
施工图设计：浙江省建筑设计研究院
摄影：奥观建筑视觉、田方方

Client: CR Land
Commission: Client Brief
Construction Brief: New Construction
Location: Hangzhou, Zhejiang
Floor Area: 486,650m²
Design/Completion: 2018/2023
Design Team:
Lead Architects: Shawn CHEUNG, HAN Zhongqiang
Architecture: WANG Zhongjie, HU Yang, HU Yifan, XUE Yihao, LI Yang, LIU Zhuoxing, JIA Gaosong, LIU Zekun, FAN Siqi, SUN Hanxiao, YU Zhenhui, JIANG Pengfei, GUO Yin
Structure: HUANG Weizhi, ZHUANG Xinlu, YAN Zhiwei, DONG Shengmin
MEP: GE Lingke, GU Liqin, SHOU Guang, NAN Xu, LIU Yuan
Collaborators:
Planning: Kohn Pedersen Fox Associates
Landscape: James Corner Field Operations, LANDAU Design
Construction Drawing: ZIAD
Images: AOGVISION, Studio FF

绿城外滩兰庭
Greentown the Bund Garden

业主：绿城中国
委托方式：直接委托
建设需求：新建
项目地点：上海市黄浦区
建筑面积：100,000m²
设计 / 竣工：2019/—
设计团队：
项目负责：陆皓、徐琦
建筑设计师：窦志国、杨冰宇、韦栋安、黄里达、戴嘉熙、余东波、张涛、王凯华
结构工程师：严志威
设备工程师：叶金元、葛令科、何骞
合作单位：
室内设计：卡莱尔设计工作室、CCD 郑中设计事务所、蓝城联合设计公司
景观设计：GTS 蓝颂设计集团
施工图设计：天华

Client: Greentown China
Commission: Client Brief
Construction Brief: New Construction
Location: Huangpu, Shanghai
Floor Area: 100,000m²
Design/Completion: 2019/-
Design Team:
Lead Architects: LU Hao, XU Qi
Architecture: DOU Zhiguo, YANG Bingyu, WEI Dong'an, HUANG Lida, DAI Jiaxi, XU Dongbo, ZHANG Tao, WANG Kaihua
Structure: YAN Zhiwei
MEP: YE Jinyuan, GE Lingke, HE Qian
Collaborators:
Interior: Carlisle Design Studio, Cheng Chung Design, Bluetown Design Studio
Landscape: GTS Lansong Design Group
Construction Drawing: TIANHUA

融创滨江杭源御潮府
Sunac Binjiang Imperial Mansion

业主：滨江集团、融创中国
项目来源：直接委托
建设需求：新建
项目地点：浙江省杭州市
建筑面积：172,700m²
设计 / 竣工：2018/2022
设计团队：
项目负责：张晓晓、于军贵
建筑设计师：林昌炎、杨晨恺、刘卓星、陈浩杰、方雁容、谢冕
设备工程师：刘源、周益林、周伟明
合作单位：
室内设计：丹健国际、CCD
景观设计：JTL
施工图设计：浙江省工业设计研究院有限公司
摄影：此间建筑摄影

Client: Binjiang, Sunac
Commission: Client Brief
Construction Brief: New Construction
Location: Hangzhou, Zhejiang
Floor Area: 172,700m²
Design/Completion: 2018/2022
Design Team:
Lead Architects: Shawn CHEUNG, YU Jungui
Architecture: LIN Changyan, YANG Chenkai, LIU Zhuoxing, CHEN Haojie, FANG Yanrong, XIE Mian
MEP: LIU Yuan, ZHOU Yilin, ZHOU Weiming
Collaborators:
Interior: DIA, CCD
Landscape: JTL
Construction Drawing: Zhejiang Industry Design & Research Institute Co., Ltd
Images: IN BETWEEN

绿城春风金沙
Greentown Hangzhou Lakeside Mansion

业主：绿城中国
委托方式：直接委托
建设需求：新建
项目地点：浙江省杭州市
建筑面积：239,800m²
设计 / 竣工：2019/2023
设计团队：
项目负责：陆皓、梁卓敏
建筑设计师：宓芬、庄雪松、涂兆云、祁雅菁、秦阗怡、熊鑫、俞俊楠、毛旃煜
结构工程师：徐浩祥、柴磊、钟锡铭、吴茂铭、赵凯龙、周建成、叶鑫
设备工程师：王文胜、陈舟舟、彭迎云、高利强、褚福华、程磊、魏山山、南旭、叶金元、张雪祁、张毓、赵睿、高涵、时云强、钱列东
合作单位：
室内设计：翰衡设计、BuregaFarnell
景观设计：GTS 蓝颂设计集团
摄影：RudyKu

Client: Greentown China
Commission: Client Brief
Construction Brief: New Construction
Location: Hangzhou, Zhejiang
Floor Area: 239,800m²
Design/Completion: 2019/2023
Design Team:
Lead Architects: LU Hao, LIANG Zhuomin
Architecture: MI Fen, ZHUANG Xuesong, TU Zhaoyun, QI Yajing, QIN Tianyi, XIONG Xin, YU Junnan, MAO Zhanyu
Structure: XU Haoxiang, CHAI Lei, ZHONG Ximing, WU Maoming, ZHAO Kailong, ZHOU Jiancheng, YE Xin
MEP: WANG Wensheng, CHEN Zhouzhou, PENG Yingyun, GAO Liqiang, YANG Fuhua, CHENG Lei, WEI Shanshan, NAN Xu, YE Jinyuan, ZHANG Xueqi, ZHANG Yu, ZHAO Rui, GAO Han, SHI Yunqiang, QIAN Liedong
Collaborators:
Interior: SWS Group, BuregaFarnell
Landscape: GTS Lansong Design Group
Images: RudyKu

华润武汉瑞府
CR Land Wuhan Park Lane Mansion

业主：华润置地
委托方式：招投标
建设需求：新建
项目地点：湖北省武汉市
建筑面积：298,900m²
设计 / 竣工：2020/—
设计团队：
项目负责：张晓晓
建筑设计师：王忠杰、张文涛、钱途、汪进、陈浩杰、汪杨帆、薛乙浩、翁飘飘、叶霞明、李起航
结构工程师：庄新炉
设备工程师：葛令科、刘源、王超伟
合作单位：
室内设计：万景百年
景观设计：加特林景观
施工图设计：正华设计
摄影：此间建筑摄影

Client: CR Land
Commission: Tender
Construction Brief: New Construction
Location: Wuhan, Hubei
Floor Area: 298,900m²
Design/Completion: 2020/-
Design Team:
Lead Architect: Shawn CHEUNG
Architecture: WANG Zhongjie, ZHANG Wentao, QIAN Tu, WANG Jin, CHEN Haojie, WANG Yangfan, XUE Yihao, WENG Piaopiao, YE Xiaming, LI Qihang
Structure: ZHUANG Xinlu
MEP: GE Lingke, LIU Yuan, WANG Chaowei
Collaborators:
Interior: InterScape Design Associates
Landscape: JTL Studio
Construction Drawing: ZHDI
Images: IN BETWEEN

蓝城陶然里
Bluetown the Kidult

业主：蓝城集团
委托方式：直接委托
建设需求：新建
项目地点：浙江省杭州市
建筑面积：130,000m²
设计 / 竣工：2018/2023
设计团队：
项目负责：张迅
建筑设计师：胡培、汪进、戴璐、陈林冰、方言智、童中拓、黄敏、余偲聪、董安悦
结构工程师：徐浩祥、柴磊、林逸风、陈欢欢、于悦、钟奇、吴强峰、金晓东、钟锡铭
设备工程师：王文胜、陈舟舟、彭迎云、王兆星、施昱展、葛令科、梅玉龙、周伟明、魏山山、胡一东、叶金元、寿广、张雪祁、张毓、赵睿、赵志铭、刘丽芳、钱列东
合作设计：
室内设计：紫香舸、蓝城装饰
景观设计：LANDAU 朗道国际设计
摄影：goa 大象设计

Client: Bluetown Group
Commission: Client Brief
Construction Brief: New Construction
Location: Hangzhou, Zhejiang
Floor Area: 130,000m²
Design/Completion: 2018/2023
Design Team:
Lead Architect: ZHANG Xun
Architecture: HU Pei, WANG Jin, DAI Lu, CHEN Linbing, FANG Yanzhi, TONG Zhongtuo, HUANG Min, YU Sicong, DONG Anyue
Structure: XU Haoxiang, CHAI Lei, LIN Yifeng, CHEN Huanhuan, YU Yue, ZHONG Qi, WU Qiangfeng, JIN Xiaodong, ZHONG Ximing
MEP: WANG Wensheng, CHEN Zhouzhou, PENG Yingyun, WANG Zhaoxing, SHI Yuzhan, GE Lingke, MEI Yulong, ZHOU Weiming, WEI Shanshan, HU Yidong, YE Jinyuan, SHOU Guang, ZHANG Xueqi, ZHANG Yu, ZHAO Rui, ZHAO Zhiming, LIU Lifang, QIAN Liedong
Collaborators:
Interior: Purple's Design, BTDECORATION
Landscape: LANDAU Design
Images: GOA

绿城空中院墅
Greentown Sky Villa

业主：绿城中国
项目来源：直接委托
建设需求：新建
项目地点：浙江省湖州市
建筑面积：1,600m²
设计 / 竣工：2020/2022
设计团队：
项目负责：张晓晓、于军贵
建筑设计师：林昌炎、申童、汪进
结构工程师：黄伟志、庄新炉、陈聪
设备工程师：寿广、陈文卉、南旭、叶金元、葛令科、黄路、曾杰、任庆军、侯会芬、张雪祁
合作单位：
室内设计：杭州形意内建筑设计有限公司
景观设计：浙江蓝颂园林景观设计集团有限公司
摄影：此间建筑摄影、奥观建筑视觉、绿城中国

Client: Greentown China
Commission: Client Brief
Construction Brief: New Construction
Location: Huzhou, Zhejiang
Floor Area: 1,600m²
Design/Completion: 2020/2022
Design Team:
Lead Architect: Shawn CHEUNG, YU Jungui
Architecture: LIN Changyan, SHEN Tong, WANG Jin
Structure: HUANG Weizhi, ZHUANG Xinlu, CHEN Cong
MEP: SHOU Guang, CHEN Wenhui, NAN Xu, YE Jinyuan, GE Lingke, HUANG Lu, ZENG Jie, REN Qingjun, HOU Huifen, ZHANG Xueqi
Collaborators:
Interior: Interior Architecture Studio
Landscape: GTS Lansong Design Group
Images: IN BETWEEN, AOGVISION, Greentown China

阿丽拉乌镇
Alila Wuzhen

业主：雅达国际
委托方式：直接委托
建设需求：新建
项目地点：浙江省嘉兴市
建筑面积：25,000m²
设计 / 竣工：2014/2018
设计团队：
项目负责：陆皓、张迅
建筑设计师：李政、陈威、陈致浩、鲁华、马佳、孟德星、裘敏
室内设计师：林琳赟、李政、诸双、戴亮亮、徐森强、何志盛
结构工程师：包凤、王琦、柴磊、赵晨、冯小生、王雪涛
设备工程师：寿广、庄少阳、南旭、王俊、周伟明、吴金祥、李星、赵志铭、钱列东、曾杰、詹鹏举、朱耀娟、孙中南、梅玉龙
合作单位：
景观设计：张唐景观
照明顾问：BPI 碧谱照明设计
酒店管理：凯悦酒店集团
摄影：是然建筑摄影

Client: YADA International
Commission: Client Brief
Construction Brief: New Construction
Location: Jiaxing, Zhejiang
Floor Area: 25,000m²
Design/Completion: 2014/2018
Design Team:
Lead Architects: LU Hao, ZHANG Xun
Architecture: LI Zheng, CHEN Wei, CHEN Zhihao, LU Hua, MA Jia, MENG Dexing, QIU Min
Interior: LIN Linyun, LI Zheng, ZHU Shuang, DAI Liangliang, XU Senqiang, HE Zhisheng
Structure: BAO Feng, WANG Qi, CHAI Lei, ZHAO Chen, FENG Xiaosheng, WANG Xuetao
MEP: SHOU Guang, ZHUANG Shaoyang, NAN Xu, WANG Jun, ZHOU Weiming, WU Jinxiang, LI Cheng, ZHAO Zhiming, QIAN Liedong, ZENG Jie, ZHAN Pengju, ZHU Yaojuan, SUN Zhongnan, MEI Yulong
Collaborators:
Landscape: Z+T Studio
Lighting: Brandston Partnership Inc.
Hotel Management: Hyatt Hotels Corporation
Images: SCHRAN

木守西溪
Muh Shoou Xixi

业主：木守世业
委托方式：直接委托
建设需求：改造
项目地点：浙江省杭州市
建筑面积：7,000m²
设计／竣工：2015/2018
设计团队：
项目负责：张晓晓
建筑设计师：杨必龙、蒋鹏飞、王旭红、王晨曦
室内设计师：任丽娜、杨必龙、毛瑜玫、周佳祺
结构工程师：师建伟、俞洪
设备工程师：曾杰、王兆星、梅玉龙、徐幸、寿广、庄少阳、陈文卉、任庆军、陈梦洁、钱列东
合作单位：
室内设计：朗图设计、BOB 陈飞波设计事务所
景观设计：张唐景观
酒店管理：木守世业
摄影：泠城摄影工作室、三风

Client: Muh Shoou Shiye
Commission: Client Brief
Construction Brief: Renovation
Location: Hangzhou, Zhejiang
Floor Area: 7,000m²
Design/Completion: 2015/2018
Design Team:
Lead Architect: Shawn CHEUNG
Architecture: YANG Bilong, JIANG Pengfei, WANG Xuhong, WANG Chenxi
Interior: REN Lina, YANG Bilong, MAO Yumei, ZHOU Jiaqi
Structure: SHI Jianwei, YU Hong
MEP: ZENG Jie, WANG Zhaoxing, MEI Yulong, XU Xing, SHOU Guang, ZHUANG Shaoyang, CHEN Wenhui, REN Qingjun, CHEN Mengjie, QIAN Liedong
Collaborators:
Interior: Naked Design Office, BOBCHEN Design Office
Landscape: Z+T Studio
Hotel Management: Muh Shoou Shiye
Images: SHIROMIO Studio, Three wind

杭州远洋凯宾斯基酒店
Kempinski Hotel Hangzhou

业主：远洋集团
委托方式：直接委托
建设需求：新建
项目地点：浙江省杭州市
建筑面积：60,000m²（地上）
设计／竣工：2012/2019
设计团队：
项目负责：凌建
建筑设计师：蒋嘉菲、李洪、刘大可、王旭红、俞培柱
结构工程师：胡凌华、庄新炉、金明彦、严志威、敖国胜
设备工程师：曾菲、王俊、叶金元、寿广、张雪祁、李程、赵志铭、张毓、郑铭、钱列东
合作单位：
施工图设计：汉嘉设计（地下室）
室内设计：ACID
景观设计：艾奕康
酒店管理：凯宾斯基酒店
摄影：泠城摄影工作室、一乘摄影

Client: Sino-Ocean Group
Commission: Client Brief
Construction Brief: New Construction
Location: Hangzhou, Zhejiang
Floor Area: 60,000m² (aboveground)
Design/Completion: 2012/2019
Design Team:
Lead Architect: LING Jian
Architecture: JIANG Jiafei, LI Hong, LIU Dake, WANG Xuhong, YU Peizhu
Structure: HU Linghua, ZHUANG Xinlu, JIN Mingyan, YAN Zhiwei, AO Guosheng
MEP: ZENG Jie, ZHU Yaojuan, YANG Fuhua, MEI Yulong, ZHOU Weiming, ZENG Fei, WANG Jun, YE Jinyuan, SHOU Guang, ZHANG Xueqi, LI Cheng, ZHAO Zhiming, ZHANG Yu, ZHENG Ming, QIAN Liedong
Collaborators:
Construction Drawing: Hanjia Design (basement)
Interior: Avalon Collective Interior Design
Landscape: AECOM
Hotel Management: Kempinski Hotel
Images: SHIROMIO Studio, Yicheng Studio

湘湖逍遥庄园
Xianghu Xiaoyao Manor

业主：湘湖逍遥有限公司
委托方式：直接委托
建设需求：新建
项目地点：浙江省杭州市
建筑面积：126,700m²
设计／竣工：2015/2019
设计团队：
项目负责：何兼、陈斌鑫
建筑设计师：吕焕政、谷裕、马惟略、刘宇澄、许晋朝、蒋经军、盛蓓蓓、张韵、李政、权晓、郑文庚、王炯巍、陈海冬、孙璐、汪海莹、祝容、刘天宇、叶俊
景观设计师：侯冬炜、袁美云、张颖、薛天炜、赵犖珅
结构工程师：包凤、徐浩祥、杜攀峰、盛建康、崔碧琪、张帆、钟锡铭、陈欢欢、高俊、董忆夏、蒋莹、周孝志、李宏强
设备工程师：程磊、江漪波、南旭、叶金元、邓雅琼、寿广、任庆军、王倩雯、何骋宇、李程、刘源、郑铭、钱列东、赵哲、花勇刚、王文胜、彭迎云、王晓卉、黄路、徐丽、梅玉龙
合作单位：
室内设计：P49 Deesign、HBA
景观设计：苏州园林设计院
摄影：goa 大象设计

Client: Xianghu Xiaoyao Co., Ltd.
Commission: Client Brief
Construction Brief: New Construction
Location: Hangzhou, Zhejiang
Floor Area: 126,700m²
Design/Completion: 2015/2019
Design Team:
Lead Architects: HE Jian, CHEN Binxin
Architecture: LYU Huanzheng, GU Yu, MA Weilue, LIU Yucheng, XU Jinchao, JIANG Jingjun, SHENG Beibei, ZHANG Yun, LI Zheng, QUAN Xiao, ZHENG Wenkang, WANG Jiongwei, CHEN Haidong, SUN Lu, WANG Haiying, ZHU Rong, LIU Tianyu, YE Jun
Landscape: HOU Dongwei, YUAN Meiyun, ZHANG Ying, XUE Tianwei, ZHAO Jianshen
Structure: BAO Feng, XU Haoxiang, DU Panfeng, SHENG Jiankang, CUI Biqi, ZHANG Fan, ZHONG Ximing, CHEN Huanhuan, GAO Jun, DONG Yixia, JIANG Ying, ZHOU Xiaozhi, LI Hongqiang
MEP: CHENG Lei, JIANG Yibo, NAN Xu, YE Jinyuan, DENG Yaqiong, SHOU Guang, REN Qingjun, WANG Qianwen, HE Chengyu, LI Cheng, LIU Yuan, ZHENG Ming, QIAN Liedong, ZHAO Zhe, HUA Yonggang, WANG Wensheng, PENG Yingyun, WANG Xiaohui, HUANG Lu, XU Li, MEI Yulong
Collaborators:
Interior: P49 Deesign, Hirsch Bedner Associates
Landscape: Suzhou Institute of Landscape Architecture Design
Images: GOA

苏州狮山悦榕庄
Banyan Tree Suzhou Shishan

业主：仁恒置地、苏高新集团
委托方式：直接委托
建设需求：新建
项目地点：江苏省苏州市
建筑面积：55,400m²
设计／竣工：2020/—
设计团队：
项目负责：张迅、陈娴
建筑设计师：江昊、林发光、杜皓月、王心恬、杨淑婷
结构设计师：徐浩祥、柴磊、王晓君、叶鑫
设备设计师：任庆军、葛令科、南旭
合作单位：
室内设计：HBA、金螳螂
景观设计：诗加达、杭州现代设计工程有限公司
施工图设计：苏州建设（集团）规划建筑设计院
酒店管理：悦榕庄

Client: Yanlord Land, SND Group
Commission: Client Brief
Construction Brief: New Construction
Location: Suzhou, Jiangsu
Floor Area: 55,400m²
Design/Completion: 2020/-
Design Team:
Lead Architect: ZHANG Xun, CHEN Xian
Architecture: JIANG Hao, LIN Faguang, DU Haoyue, WANG Xintian, YANG Shuting
Structure: XU Haoxiang, CHAI Lei, WANG Xiaojun, YE Xin
MEP: REN Qingjun, GE Lingke, NAN Xu
Collaborators:
Interior: Hirsch Bedner Associates, Gold Mantis
Landscape: CICADA, MEDG
Construction Drawing: SCG Architecture & Urban Planning
Hotel Management: BANYAN TREE

德清莫干山洲际酒店
InterContinental Deqing Moganshan

业主：德清县博信旅游开发有限公司
委托方式：招投标
建设需求：新建
项目地点：浙江省湖州市
建筑面积：53,000m²
设计 / 竣工：2020/—
设计团队：
项目负责：田钰、胡晨芳
建筑设计师：林华通、孙周强、丁培峰、钱程、徐龙奇、赵恩帅、吴寒蕴、刘筱珉、孙雅贤
室内设计师：林琳赟、诸双、胡文涛、郭思聪、袁凯、黄牧舟、徐怀忱
景观设计师：侯冬炜、姜欢、邢益楠、杨怡然、郑涛、张颖、朱静
结构工程师：黄伟志、庄新炉、龚铭、朱芳、任庆阳、俞洪、吴强峰、金鑫、莫佳杰
设备工程师：钱列东、张雪祁、任庆军、何骞、赵志铭、侯会芬、张毓、杨福华、黄池钧、彭迎云、施昱展、葛令科、曾杰、程磊、卢琦、孟娜、寿广、叶金元
合作单位：
施工图设计：西城工程设计集团有限公司
酒店管理：洲际酒店集团

Client: Deqing County Boxin Tourism Development Co., Ltd.
Commission: Tender
Construction Brief: New Construction
Location: Huzhou, Zhejiang
Floor Area: 53,000m²
Design/Completion: 2020/-
Design Team:
Lead Architect: TIAN Yu, HU Chenfang
Architecture: LIN Huatong, SUN Zhouqiang, DING Peifeng, QIAN Cheng, XU Longqi, ZHAO Enshuai, WU Hanyun, LIU Xiaomin, SUN Yaxian
Interior: LIN Linyun, ZHU Shuang, HU Wentao, GUO Sicong, YUAN Kai, HUANG Muzhou, XU Huaichen
Landscape: HOU Dongwei, JIANG Huan, XING Yinan, YANG Yiran, ZHENG Tao, ZHANG Ying, ZHU Jing
Structure: HUANG Weizhi, ZHUANG Xinlu, GONG Ming, ZHU Fang, REN Qingyang, YU Hong, WU Qiangfeng, JIN Xin, MO Jiajie
MEP: QIAN Liedong, ZHANG Xueqi, REN Qingjun, HE Qian, ZHAO Zhiming, HOU Huifen, ZHANG Yu, YANG Fuhua, HUANG Chijun, PENG Yingyun, SHI Yuzhan, GE Lingke, ZENG Jie, CHENG Lei, LU Qi, MENG Na, SHOU Guang, YE Jinyuan
Collaborators:
Construction Drawing: Xicheng Engineering Design Group Co., Ltd.
Hotel Management: InterContinental Hotels Group

湘湖陈家埠酒店
Xianghu Chenjiabu Hotel

业主：湘湖逍遥有限公司
委托方式：直接委托
建设需求：新建
项目地点：浙江省杭州市
建筑面积：58,000m²
设计 / 竣工：2015/2023
设计团队：
项目负责：何兼、田钰
建筑设计师：胡晨芳、朱银杰、林虎、柴页新、杜立明
结构工程师：黄伟志、詹伟良、王立才、俞洪、盛建康、汪卓红、周孝志、方美平、洪飞、高俊、马永宏
设备工程师：王超伟、邓雅琼、魏山山、王俊、寿广、叶金元、任庆军、侯会芬、王倩雯、何骋宇、张雪祁、葛令科、谷立芹、李翔、王兆星、黄路、曾杰
合作单位：
室内设计：MOD 穆德设计
景观设计：M.A.O.
摄影：goa 大象设计

Client: Xianghu Xiaoyao Co., Ltd.
Commission: Client Brief
Construction Brief: New Construction
Location: Hangzhou, Zhejiang
Floor Area: 58,000m²
Design/Completion: 2015/2023
Design Team:
Lead Architects: HE Jian, TIAN Yu
Architecture: HU Chenfang, ZHU Yinjie, LIN Hu, CHAI Yexin, DU Liming
Structure: HUANG Weizhi, ZHAN Weiliang, WANG Licai, YU Hong, SHENG Jiankang, WANG Zhuohong, ZHOU Xiaozhi, FANG Meiping, HONG Fei, GAO Jun, MA Yonghong
MEP: WANG Chaowei, DENG Yaqiong, WEI Shanshan, WANG Jun, SHOU Guang, YE Jinyuan, REN Qingjun, HOU Huifen, WANG Qianwen, HE Chengyu, ZHANG Xueqi, GE Lingke, GU Liqin, LI Xiang, WANG Zhaoxing, HUANG Lu, ZENG Jie
Collaborators:
Interior: MOD Interior Design
Landscape: Masters' Architectural Office
Images: GOA

青岛藏马山酒店
Qingdao Cangmashan Hotel

业主：融创中国
委托方式：直接委托
建设需求：新建
项目地点：山东省青岛市
建筑面积：20,260m²
设计 / 竣工：2018/—
设计团队：
项目负责：陆皓、徐琦
建筑设计师：杜立明、郝睿敏、黄里达、张富强、花子杰、张蓓艳、沈强、戴嘉熙
结构工程师：师建伟、金明彦、周建成、叶怀晨、严志威
设备工程师：叶金元、王超伟、屠兴灿、王雅迪、寿广、任庆军、张雪祁、刘源、王俊涛、徐幸、姚银杰、葛令科、曾杰、黄路、谷立芹
合作单位：
室内设计：水平线设计
景观设计：TOPO

Client: Sunac China
Commission: Client Brief
Construction Brief: New Construction
Location: Qingdao, Shandong
Floor Area: 20,260m²
Design/Completion: 2018/-
Design Team:
Lead Architects: LU Hao, XU Qi
Architecture: DU Liming, HAO Ruimin, HUANG Lida, ZHAN Fuqiang, HUA Zijie, ZHANG Beiyan, SHEN Qiang, DAI Jiaxi
Structure: SHI Jianwei, JIN Mingyan, ZHOU Jiancheng, YE Huaichen, YAN Zhiwei
MEP: YE Jinyuan, WANG Chaowei, TU Xingcan, WANG Yadi, SHOU Guang, REN Qingjun, ZHANG Xueqi, LIU Yuan, WANG Juntao, XU Xing, YAO Yinjie, GE Lingke, ZENG Jie, HUANG Lu, GU Liqin
Collaborators:
Interior: Horizontal Design
Landscape: TOPO

既下山大同
SUNYATA Hotel Datong

业主：华夏江鸿（大同）文化旅游开发有限公司
委托方式：直接委托
建设需求：新建 + 改造
项目地点：山西省大同市
建筑面积：9,800m²
设计 / 竣工：2021/—
设计团队：
项目负责：张晓晓
建筑设计师：杨必龙、王国兴、韦晨璐、林梅娟、周雨馨、黄楚阳
结构工程师：谢忠威
设备工程师：杨福华、候会芬、孟娜
合作单位：
室内设计：CCD 郑中设计事务所
施工图设计：BCG 联创机构
酒店管理：既下山

Client: Huaxia Jianghong (Datong) Cultural Tourism Development Co., Ltd.
Commission: Client Brief
Construction Brief: New Construction & Renovation
Location: Datong, Shanxi
Floor Area: 9,800m²
Design/Completion: 2021/-
Design Team:
Lead Architect: Shawn CHEUNG
Architecture: YANG Bilong, WANG Guoxing, WEI Chenlu, LIN Meijuan, ZHOU Yuxin, HUANG Chuyang
Structure: XIE Zhongwei
MEP: YANG Fuhua, HOU Huifen, MENG Na
Collaborators:
Interior: Cheng Chung Design
Construction Drawing: Brother Cooperation Group
Hotel Management: SUNYATA Hotels

阳羡溪山
Yangxian Landscape

业主：雅达国际
委托方式：直接委托
建设需求：新建
项目地点：江苏省无锡市
建筑面积：635,000m²
设计/竣工：2017/—
设计团队：
项目负责：陆皓、陈斌鑫、徐琦、王彦
建筑设计师：袁波、刘天宇、吕焕政、谷裕、祝志根、陈林、石逸雄、赖雨诗、马惟略、徐艳、岳海旭、刘宇澄、郑文康、李振、卢耿聪、李富贵、刘广荣、叶李洁、蒋经军、吴岳肄、施旗、陈娴、杜立明、欧阳之曦、李亚萍、周天宇、尹丽荣、钱杰、薛华、邓蔚超、刘宇、夏丽丽、徐美锋、张韵、肖杰、何儒迪、孟戍、黄瑞国、陈静怡、王冰卿、蒋骁闻、涂兆云、郭华、周李飞、尹建博、汪进、邢益楠、王羽、关伟超、李力、吴若晨、钟铖、刘哲圣、吴宁宁、陆薇薇、张蓓艳、辛博文、邵立文、周天宇、黄星达、杨冰宇、熊志丹、高力、吴若晨、王心恬
结构工程师：黄伟志、庄新炉、叶怀晨、洪飞、谢忠威、王立才、陈聪、崔碧琪、方美平
设备工程师：任庆军、陈梦洁、王倩雯、王客、何骋宇、王俊涛、刘源、钱冬东、张毓、张雪祁、曾杰、黄琦、王兆星、黄路、葛令科、杨福华、庄少阳、南旭、胡一东、魏山山、陈文卉、叶金元、程磊、王超伟、寿广、徐幸、吕小斌、盛梦乡、时云强、高涵、徐丽、卢琦
景观设计师：侯冬炜、张颖、郑涛、黄佳英、薛天炜、袁美云、卓百会、赵举坤
BIM 工程师：李潇乐、赵霏霏、王凯烽、徐君豪、王彦韬、袁一、姚远、赵烁
合作单位：
景观设计：绿城风景
摄影：goa 大象设计、泠城摄影工作室

Client: YADA International
Commission: Client Brief
Construction Brief: New Construction
Location: Wuxi, Jiangsu
Floor Area: 635,000m²
Design/Completion: 2017/-
Design Team:
Lead Architects: LU Hao, CHEN Binxin, XU Qi, WANG Yan
Architecture: YUAN Bo, LIU Tianyu, LYU Huanzheng, GU Yu, ZHU Zhigen, CHEN Lin, SHI Yixiong, LAI Yushi, MA Weilue, XU Yan, YUE Haixu, LIU Yucheng, ZHENG Wenkang, LI Zhen, LU Gengcong, LI Fugui, LIU Guangrong, YE Lijie, JIANG Jingjun, WU Yuexiao, SHI Qi, CHEN Xian, DU Liming, OUYANG Zhixi, LI Yaping, ZHOU Tianyu, YIN Lirong, QIAN Jie, XUE Hua, DENG Weichao, LIU Yu, XIA Lili, XU Meifeng, ZHANG Yun, XIAO Jie, HE Rudi, MENG Cheng, HUANG Ruiguo, CHEN Jingyi, WANG Bingqing, JIANG Xiaowen, TU Zhaoyun, GUO Hua, ZHOU Lifei, YIN Jianbo, WANG Jin, XING Yinan, WANG Yu, GUAN Weichao, LI Li, WU Ruochen, ZHONG Cheng, LIU Zhesheng, WU Ningning, LU Weiwei, ZHANG Beiyan, XIN Bowen, SHAO Liwen, ZHOU Tianyu, HUANG Lida, YANG Bingyu, XIONG Zhidan, GAO Li, WU Ruochen, WANG Xintian
Structure: HUANG Weizhi, ZHUANG Xinlu, YE Huaichen, HONG Fei, XIE Zhongwei, WANG Licai, CHEN Cong, CUI Biqi, FANG Meiping
MEP: REN Qingjun, CHEN Mengjie, WANG Qianwen, WANG Ke, HE Chengyu, WANG Juntao, LIU Yuan, QIAN Liedong, ZHANG Yu, ZHANG Xueqi, ZENG Jie, HUANG Qi, WANG Zhaoxing, HUANG Lu, GE Lingke, YANG Fuhua, ZHUANG Shaoyang, NAN Xu, HU Yidong, WEI Shanshan, CHEN Wenhui, YE Jinyuan, CHENG Lei, WANG Chaowei, SHOU Guang, XU Xing, LYU Xiaobin, SHENG Mengyun, SHI Yunqiang, GAO Han, XU Li, LU Qi
Landscape: HOU Dongwei, ZHANG Ying, ZHENG Tao, HUANG Jiaying, XUE Tianwei, YUAN Meiyun, ZHUO Baihui, ZHAO Jianshen
BIM: LI Xiaole, ZHAO Feifei, WANG Kaifeng, XU Junhao, WANG Yantao, YUAN Yi, YAO Yuan, ZHAO Shuo
Collaborators:
Landscape: Green Townscape
Images: GOA, SHIROMIO Studio

曹山未来城古桥水镇
CaoShan Future City Guqiao Water Town

业主：融创中国
项目来源：直接委托
建设需求：新建
项目地点：江苏省常州市
建筑面积：171,000m²
设计/竣工：2019/—
设计团队：
项目负责：何峻、陈斌鑫、陈伟
建筑设计师：袁波、赵栋、张华杰、李梦萌、周丁一、王家成、黄韬、李振、刘广荣
结构工程师：杨洋、何亮、龚铭
设备工程师：王超伟、高利强、张毓
合作单位：
景观设计：安道设计
施工图设计：江苏筑原建筑设计有限公司
摄影：融创中国

Client: SUNAC China
Commission: Client Brief
Construction Brief: New Construction
Location: Liyang, Jiangsu
Floor Area: 171,000m²
Design/Completion: 2019/-
Design Team:
Lead Architects: HE Jun, CHEN Binxin, CHEN Wei
Architecture: YUAN Bo, ZHAO Dong, ZHANG Huajie, LI Mengmeng, ZHOU Dingyi, WANG Jiacheng, HUANG Tao, LI Zhen, LIU Guangrong
Structure: YANG Yang, HE Liang, GONG Ming
MEP: WANG Chaowei, GAO Liqiang, ZHANG Shu
Collaborators:
Landscape: Antao
Construction Drawing: Jiangsu Zhuyuan Build Design Co., Ltd
Images: SUNAC China

张謇故里小镇柳西半街
Jianli Town Liuxiban Street

业主：中南集团
委托方式：直接委托
建设需求：新建
项目地点：江苏省南通市
建筑面积：14,600m²
设计/竣工：2019/2021
设计团队：
项目负责：王彦
建筑设计师：郑静云、殷长伟、夏斐、刘杜娟、彭书勉、徐晨歆、吴超楠、苏子茗、李夫龙、陈振
结构工程师：何亮
设备工程师：何骞、曾杰、周伟明
合作单位：
室内设计：夏谷暑雨
景观设计：朗道设计
施工图设计：江苏华源建筑设计研究院股份有限公司
摄影：此间建筑摄影

Client: YADA International
Client: Zhongnan Group
Commission: Client Brief
Construction Brief: New Construction
Location: Nantong, Jiangsu
Floor Area: 14,600m²
Design/Completion: 2019/2021
Design Team:
Lead Architects: WANG Yan
Architecture: ZHENG Jingyuan, YING Changwei, XIA Fei, LIU Dujuan, PENG Shumian, XU Chenyin, WU Chaonan, SU Ziming, LI Fulong, CHEN Zhen
Structure: HE Liang
MEP: HE Qian, ZENG Jie, ZHOU Weiming
Collaborators:
Interior: XIA GU SHU YU
Landscape: LANDAU Design
Construction Drawing: Huayuan Architectural Design & Research Institute co., Ltd.
Images: IN BETWEEN

曲水善湾乡村振兴示范区
Qushui Shanwan Rural Revitalization Demonstration Area

业主：苏州汾湖投资集团有限公司
委托方式：直接委托
建设需求：新建
项目地点：江苏省苏州市
建筑面积：1,300m²
设计/竣工：2020/2022
设计团队：
项目负责：陈斌鑫
建筑设计师：叶李洁、杜立明、徐艳、施旗、黄瑞国、魏瑞环、刘天宇、袁波、谷裕、刘安、卢文华、石逸雄
结构工程师：黄伟志、詹伟良、周建成
设备工程师：刘源、张雪祁、任庆军、彭迎云、曾杰、王文胜、孟娜、叶金元、程磊
合作单位：
室内设计：益善堂设计、万境设计
景观设计：利恩设计
运营顾问：苏州蓝城文旅有限公司
摄影：此间建筑摄影

Client: Suzhou Fenhu Investment Group Co., Ltd.
Commission: Client Brief
Construction Brief: New Construction
Location: Suzhou, Jiangsu
Floor Area: 1,300m²
Design/Completion: 2020/2022
Design Team:
Lead Architect: CHEN Binxin
Architecture: YE Lijie, DU Liming, XU Yan, SHI Qi, HUANG Ruiguo, WEI Ruihuan, LIU Tianyu, YUAN Bo, GU Yu, LIU An, LU Wenhua, SHI Yixiong
Structure: HUANG Weizhi, ZHAN Weiliang, ZHOU Jiancheng
MEP: LIU Yuan, ZHANG Xueqi, REN Qingjun, PENG Yingyun, ZENG Jie, WANG Wensheng, MENG Na, YE Jinyuan, CHENG Lei
Collaborators:
Interior: YST Design, WJ STUDIO
Landscape: 9+LEON Organization
Operation Consultant: Suzhou Bluetown Cultural Tourism Co., Ltd.
Images: IN BETWEEN

雅达剧院
Yada Theater

业主：雅达国际
委托方式：直接委托
建设需求：新建
项目地点：江苏省无锡市
建筑面积：6,000m²
设计/竣工：2019/2022
设计团队：
项目负责：陆皓、徐琦
建筑设计师：熊志丹、黄星达、杨冰宇、高利、沈强、吴若晨、王心恬、王培柱
室内设计师：林琳赞、胡文涛、戴亮亮、诸双、郭思聪、赵怡洁
景观设计师：侯冬炜、姜欢、郑涛、袁美云、黄佳英
结构工程师：黄伟志、庄新炉、谢忠威、洪飞
设备工程师：曾杰、黄琦、黄路、任庆军、王倩雯、钱冬东、南旭、陈文卉、程磊、叶金元
合作单位：
照明顾问：同济大学建筑设计研究院照明所
声学顾问：SM&W 声美华
摄影：奥观建筑视觉、goa 大象设计

Client: YADA International
Commission: Client Brief
Construction Brief: New Construction
Location: Wuxi, Jiangsu
Floor Area: 6,000m²
Design/Completion: 2019/2022
Design Team:
Lead Architects: LU Hao, XU Qi
Architecture: XIONG Zhidan, HUANG Lida, YANG Bingyu, GAO Li, SHEN Qiang, WU Ruochen, WANG Xintian, WANG Peizhu
Interior: LIN Linyun, HU Wentao, DAI Liangliang, ZHU Shuang, GUO Sicong, ZHAO Yijie
Landscape: Hou Dongwei, Jiang Huan, Zheng Tao, Yuan Meiyun, Huang Jiaying
Structure: HUANG Weizhi, ZHUANG Xinlu, XIE Zhongwei, HONG Fei
MEP: ZENG Jie, HUANG Qi, HUANG Lu, REN Qingjun, WANG Qianwen, QIAN Liedong, NAN Xu, CHEN Wenhui, CHENG Lei, YE Jinyuan
Collaborators:
Lighting: TJAD-Lighting
Acoustic: Shen Milsom & Wilke
Images: AOGVISION, GOA

沪杭高速嘉兴服务区
G60 Expressway Jiaxing Service Area

业主：浙江省交通投资集团有限公司
委托方式：招投标
建设需求：新建
项目地点：浙江省嘉兴市
建筑面积：21,800m²
设计/竣工：2020/2022
设计团队：
项目负责：何兼、袁源
建筑设计师：郭吟、李凌、李宏捷、俞翔、叶帆、蒋洒洒、潘雨禾、徐子帆
景观设计师：侯冬炜、黄佳英、郑涛
结构工程师：徐浩祥、陈森、王琦、任庆阳、陈聪、潘鹏
设备工程师：葛令科、王文胜、高利强、谷立芹、曾杰、张雪祁、任庆军、李程、何骋宇、吴金祥、寿广、程磊、叶金元
全过程工程咨询管理：沈旭凯、赵亮亮、童玥、高俊、魏民、任丽娜、王周峰、卢琳、孙杭金
合作单位：
室内设计：goa 乐空设计
摄影：奥观建筑视觉、goa 大象设计

Client: Zhejiang Communications Investment Group Co., Ltd.
Commission: Tender
Construction Brief: New Construction
Location: Jiaxing, Zhejiang
Floor Area: 21,800m²
Design/Completion: 2020/2022
Design Team:
Lead Architects: HE Jian, YUAN Yuan
Architecture: GUO Yin, LI Ling, LI Hongjie, YU Xiang, YE Fan, JIANG Sasa, PAN Yuhe, XU Zifan
Landscape: HOU Dongwei, HUANG Jiaying, ZHENG Tao
Structure: XU Haoxiang, CHEN Sen, WANG Qi, REN Qingyang, CHEN Cong, PAN Peng
MEP: GE Lingke, WANG Wensheng, GAO Liqiang, GU Liqin, ZENG Jie, ZHANG Xueqi, REN Qingjun, LI Cheng, HE Chengyu, WU Jinxiang, SHOU Guang, CHENG Lei, YE Jinyuan
Whole Process Engineering Consulting Management: SHEN Xukai, ZHAO Liangliang, TONG Yue, GAO Jun, WEI Min, REN Lina, WANG Zhoufeng, LU Ling, SUN Hangjin
Collaborators:
Interior: GOA LOKON
Images: AOGVISION, GOA

建德市文化综合体
Jiande Cultural Center

业主：建德市城市建设发展投资有限公司
委托方式：直接委托
建设需求：新建
项目地点：浙江省杭州市
建筑面积：44,900m²
设计/竣工：2013/2021
设计团队：
项目负责：凌建、陈斌鑫
建筑设计师：杜立明、陈林、谷裕、徐美峰、吴岳啸、蒋经军
结构工程师：师建伟、倪志军、龚铭、杨钦普、周孝志、周南
设备工程师：吴金祥、郑铭、陈梦洁、李文璧、李程、赵莹、黄钦鹏、梅玉龙、王文胜、杨福华、王晓卉、周益林、花勇刚、寿广、周伟明、吴文裕、王超伟、朱应钦、胡一东
合作单位：
室内设计：国美设计（图书馆）；龙邦建设（博物馆）
景观设计：绿城风景
摄影：夏至、goa 大象设计

Client: Jiande City Contruction Development Co., Ltd.
Commission: Client Brief
Construction Brief: New Construction
Location: Hangzhou, Zhejiang
Floor Area: 44,900m²
Design/Completion: 2013/2021
Design Team:
Lead Architects: LING Jian, CHEN Binxin
Architecture: DU Liming, CHEN Lin, GU Yu, XU Meifeng, WU Yuexiao, JIANG Jingjun
Structure: SHI Jianwei, NI Zhijun, GONG Ming, YANG Qinpu, ZHOU Xiaozhi, ZHOU Nan
MEP: WU Jinxiang, ZHENG Ming, CHEN Mengjie, LI Wenbi, LI Cheng, ZHAO Ying, HUANG Qinpeng, MEI Yulong, WANG Wensheng, YANG Fuhua, WANG Xiaohui, ZHOU Yilin, HUA Yonggang, SHOU Guang, ZHOU Weiming, WU Wenyu, WANG Chaowei, ZHU Yingqin, HU Yidong
Collaborators:
Interior: GMAID (library); LongBang Construction (museum)
Landscape: Green Townscape
Images: XIA Zhi, GOA

舟山绿城育华幼儿园
Zhoushan Greentown Yuhua Kindergarten

业主：绿城中国
委托方式：直接委托
建设需求：新建
项目地点：浙江省舟山市
建筑面积：13,500m²
设计/竣工：2018/2021
设计团队：
项目负责：张迅
建筑设计师：汪进、刘昱雪、童忠拓
景观设计师：侯冬炜、王羽、袁美云、黄佳英、郑涛
结构工程师：徐浩祥、柴磊、赵凯龙
设备工程师：毛ester华、周益林、葛令科、侯会芬、王倩雯、刘丽芳、张雪祁、周伟明、朱应钦、王超伟、叶金元
合作单位：
室内设计：禾泽都林
摄影：goa 大象设计

Client: Greentown China
Commission: Client Brief
Construction Brief: New Construction
Location: Zhoushan, Zhejiang
Floor Area: 13,500m²
Design/Completion: 2018/2021
Design Team:
Lead Archtiect: ZHANG Xun
Architecture: WANG Jin, LIU Yuxue, TONG Zhongtuo
Landscape: HOU Dongwei, WANG Yu, YUAN Meiyun, HUANG Jiaying, ZHENG Tao
Structure: XU Haoxiang, CHAI Lei, ZHAO Kailong
MEP: MAO Dihua, ZHOU Yilin, GE Lingke, HOU Huifen, WANG Qianwen, LIU Lifang, ZHANG Xueqi, ZHOU Weiming, ZHU Yingqin, WANG Chaowei, YE Jinyuan
Collaborators:
Interior Design: HESOMS
Images: GOA

飞鸟剧场
Earth Valley Theater

业主：江苏茗岭窑湖小镇旅游有限公司
委托方式：直接委托
建设需求：新建
项目地点：江苏省无锡市
建筑面积：9,200m²
设计/竣工：2021/—
设计团队：
项目负责：陆皓、徐琦
建筑设计师：黄里达、熊志丹、刘杜鹃、王崇宇、李静姝、李怡凝、陈振、张蓓艳、陈俊逸
结构工程师：黄伟志、庄新炉、江振鑫、吴强峰
设备工程师：曾杰、黄琦、宋晓天、葛令科、杨福华、任庆军、何骋宇、毛瑞琳、张毓、张雪祁、南旭、卢琦、王超伟、寿广、徐幸、刘佳莹
室内设计师：李扬、戴亮亮、袁凯
景观设计师：侯冬炜、张颖、李亚萍、杨怡然、邢益楠、郑涛
设计管理：沈旭凯、陈森
合作单位：
艺术顾问：跃上工作室
声学顾问：易科
环境视觉顾问：图石设计
鸟类顾问：普德赋

Client: Jiangsu Mingling Yaohu Town Tourism Co., Ltd.
Commission: Client Brief
Construction Brief: New Construction
Location: Wuxi, Jiangsu
Floor Area: 9,200m²
Design/Completion: 2021/-
Design Team:
Lead Architects: LU Hao, XU Qi
Architecture: HUANG Lida, XIONG Zhidan, LIU Dujuan, WANG Chongyu, LI Jingshu, LI Yining, CHEN Zhen, ZHANG Beiyan, CHEN Junyi
Interior: LI Yang, DAI Liangliang, YUAN Kai
Landscape: HOU Dongwei, ZHANG Ying, LI Yaping, YANG Yiran, XING Yinan, ZHENG Tao
Structure: HUANG Weizhi, ZHUANG Xinlu, JIANG Zhenxin, WU Qiangfeng
MEP: ZENG Jie, HUANG Qi, SONG Xiaotian, GE Lingke, YANG Fuhua, REN Qingjun, HE Chengyu, MAO Ruilin, ZHANG Yu, ZHANG Xueqi, NAN Xu, LU Qi, WANG Chaowei, SHOU Guang, XU Xing, LIU Jiaying
Design Management: SHEN Xukai, CHEN Sen
Collaborators:
Art Consultant: Yueshang Studio
Acoustic: EZPro
Environmental Visual Consultant: ToThree
Aviary Consultant: Puy du Fou

天台山雪乐园
Tiantaishan Snow Park

业主：绿城中国
委托方式：直接委托
建设需求：改造
项目地点：浙江省台州市
建筑面积：19,400m²
设计/竣工：2018/2021
设计团队：
项目负责：凌建、陈吉
建筑设计师：邹洁琳、陈禹男、周星宇、张弦、杜皓月、郑亚军
结构工程师：师建伟、龚铭、俞洪
设备工程师：曾杰、程磊、侯会芬
合作单位：
建筑设计：维拓设计
室内设计：优地易
景观设计：优地易
摄影：RudyKu

Client: Greentown China
Commission: Client Brief
Construction Brief: Renovation
Location: Taizhou, Zhejiang
Floor Area: 19,400m²
Design/Completion: 2018/2021
Design Team:
Lead Architect: LING Jian, CHEN Ji
Architecture: ZOU Jielin, CHEN Yunan, ZHOU Xingyu, ZHANG Xian, DU Haoyue, ZHENG Yajun
Structure: SHI Jianwei, GONG Ming, YU Hong
MEP: ZENG Jie, CHENG Lei, HOU Huifen
Collaborators:
Architecture: Victory Star
Interior: UDe
Landscape: UDe
Images: RudyKu

项目列表
Chronology

2018

绿城新疆绿城广场
新疆乌鲁木齐
居住
Greentown Xinjiang
Greentown Square
Urumqi, Xinjiang
Residential

永恒杭州之江国际商务中心
浙江杭州
办公
Yongheng Hangzhou River
International Business Center
Hangzhou, Zhejiang
Office

中地武汉中国园艺小镇
湖北武汉
文旅
Zhongdi Wuhan Chinese
Horticultural Town
Wuhan, Hubei
Cultural Tourism

百灵贵州安顺平安广场东侧住宅
贵州安顺
居住
Bailing Guizhou Anshun East Ping'an
Square Residence
Anshun, Guizhou
Residential

绿城金地武汉凤起听澜
湖北武汉
居住
Greentown & Gemdale Wuhan
Phoenix Mansion
Wuhan, Hubei
Residential

青岛藏马山酒店
山东青岛
酒店
Qingdao Cangmashan Hotel
Qingdao, Shandong
Hospitality

融创金成臻华府
浙江杭州
居住
Sunac Jincheng Zhenhua Mansion
Hangzhou, Zhejiang
Residential

融创阿朵小镇
山东青岛
文旅
Sunac A Dream A Life
Qingdao, Shandong
Cultural Tourism

绿管珠海东坑村村民活动中心
浙江杭州
城市更新
Greentown Construction
Management Zhuhai Dongkeng
Villagers Activity Center
Hangzhou, Zhejiang
Urban Renewal

九龙仓杭州天玺
浙江杭州
居住
Wharf Hangzhou Parc Royale
Hangzhou, Zhejiang
Residential

杭州地铁3号线小和山停车场上盖开发
浙江杭州
TOD
Development of Hangzhou Metro
Line 3 Xiaoheshan Parking Lot
Superstructure
Hangzhou, Zhejiang
TOD

中泰街道南湖小镇安置房
浙江杭州
居住
Zhongtai Street Nanhu Town
Resettlement Housing
Hangzhou, Zhejiang
Residential

融创金成臻蓝府
浙江杭州
居住
Sunac Jincheng Zhenlan Mansion
Hangzhou, Zhejiang
Residential

融创天津北辰东道住宅
天津
居住
Sunac Tianjin East Beichen
Road Residence
Tianjin
Residential

融创瀚海大河宸院二期
河南郑州
居住
Sunac Hanhai Dahe Chen
Institute phase II
Zhengzhou, Henan
Residential

九龙仓返湾雅苑
江苏苏州
居住
Wharf Residence Yayuan Houwan
Suzhou, Jiangsu
Residential

绿管金华塘雅苑
浙江金华
居住
Greentown Construction
Management Jinhua Tangya
Jinhua, Zhejiang
Residential

西子智慧产业园
浙江杭州
城市更新/办公/产业/商业
Xizi Wisdom Industrial Park
Hangzhou, Zhejiang
Urban Renewal/Office/Industry/
Commercial

中美低碳杭州龙井路会所
浙江杭州
酒店
Zhongmei Hangzhou Low-carbon
Longjing Road Clubhouse
Hangzhou, Zhejiang
Hospitality

顺发恒业杭州童家塘住宅
浙江杭州
居住
Shunfa Hengye Hangzhou
Tongjiatang Residence
Hangzhou, Zhejiang
Residential

融创上海森兰项目
上海
规划
Sunac Shanghai Sunland Project
Shanghai
Planning

绿城杭州西溪诚园体育公园
浙江杭州
体育
Greentown Hangzhou Xixi
Chengyuan Sports Park
Hangzhou, Zhejiang
Sports

融创哈尔滨万达城项目6#地块
黑龙江哈尔滨
居住
Sunac Harbin Wanda City
Plot 6# Residence
Harbin, Heilongjiang
Residential

融创大连御栖湖
辽宁大连
居住
Sunac Dalian Lake Palace
Dalian, Liaoning
Residential

沈阳融创城
辽宁沈阳
居住
Shenyang Sunac City
Shenyang, Liaoning
Residential

勋望小学中德校区
辽宁沈阳
教育
Xunwang Primary School
Zhongde Campus
Shenyang, Liaoning
Education

融创西安洪庆新城住宅
陕西西安
居住
Sunac Xi'an Hongqing
New City Residence
Xi'an, Shaanxi
Residential

滨江万潮星汇
浙江杭州
办公/居住
Binjiang Stellar Mansion
Hangzhou, Zhejiang
Office/Residential

东樱上海江宁路办公楼改造
上海
城市更新/办公
Dongying Shanghai Jiangning Road
Office Building Renewal
Shanghai
Urban Renewal/Office

哈尔滨融创城一期
黑龙江哈尔滨
居住
Harbin Sunac City Phase I
Harbin, Heilongjiang
Residential

百联上海北宝兴路改造
上海
城市更新/规划
Bailian Shanghai Beibaoxing
Road Renewal
Shanghai
Urban Renewal/Planning

宁波地铁5号线停车场上盖开发
浙江宁波
TOD
Development of Ningbo Metro Line
5 Parking Lot Superstructure
Ningbo, Zhejiang
TOD

绿城杭州萧山府前路住宅
浙江杭州
居住
Greentown Hangzhou Xiaoshan
Fuqian Road Residence
Hangzhou, Zhejiang
Residential

绿城常熟明月兰庭
江苏常熟
居住
Greentown Changshu Courtyard
of Bright Moon
Changshu, Jiangsu
Residential

绿城雄安容祥路国宾馆概念设计
河北保定
酒店
Greentown Xiong'an Rongxiang Road
State Guesthouse Concept Design
Baoding, Hebei
Hospitality

绿城长虹南苑
浙江杭州
居住
Greentown South Changhong
Garden
Hangzhou, Zhejiang
Residential

绿城柳州杨柳郡
广西柳州
居住
Greentown Liuzhou Yangliu County
Liuzhou, Guangxi
Residential

瑞安上海马厂路项目
上海
城市更新
Shui On Shanghai Machang
Road Project
Shanghai
Urban Renewal

绿城湖境云庐
浙江杭州
居住
Greentown Hangzhou Oriental Villa
Hangzhou, Zhejiang
Residential

重庆西政望园
重庆
居住
Chongqing Renown Garden of
Southwest University of Political
Science & Law
Chongqing
Residential

绿城杭州未来科技城沈邱路住宅
浙江杭州
居住
Greentown Hangzhou Future
Sci-Tech City Shenqiu Road Residence
Hangzhou, Zhejiang
Residential

绿城张家港港城大道住宅
江苏张家港
居住
Greentown Zhangjiagang Gangcheng
Avenue Residence
Zhangjiagang, Jiangsu
Residential

大庄山谷·乡根一叶
浙江宁波
文旅
Dazhuang Valley Hot Spring Town
Ningbo, Zhejiang
Cultural Tourism

新湖绿城启东海上明月
江苏南通
居住
Xinhu & Greentown Qidong
The Moonlit Town
Nantong, Jiangsu
Residential

融创渭南堤顶路项目
陕西渭南
商业
Sunac Weinan Diding Road Project
Weinan, Shaanxi
Commercial

无锡融创文化旅游城主题乐园商业街
江苏无锡
商业
Wuxi Sunac Cultural Tourism City
Theme Park Commercial Street
Wuxi, Jiangsu
Commercial

华润亚奥城
浙江杭州
居住
CR Land The Century City
Hangzhou, Zhejiang
Residential

融创长春御湖宸院
吉林长春
居住
Sunac Changchun Lakeview Courtyard
Changchun, Jilin
Residential

蓝城上虞运河江南里
浙江绍兴
文旅
Bluetown Shangyu Jiangnan Li
Shaoxing, Zhejiang
Cultural Tourism

蓝城海盐如意春风
浙江嘉兴
文旅
Bluetown Haiyan Ruyi Chunfeng
Jiaxing, Zhejiang
Cultural Tourism

蓝城安吉天使小镇·浅山明月
浙江湖州
居住
Bluetown Anji Town of
Angels · Qian Shan Ming Yue
Huzhou, Zhejiang
Residential

顺发恒业杭州萧山万向路办公楼
浙江杭州
办公
Shunfa Hengye Hangzhou Xiaoshan
Wanxiang Road
Office Building
Hangzhou, Zhejiang
Office

蓝城牧云谷
浙江湖州
居住
Bluetown Muyun Valley
Huzhou, Zhejiang
Residential

融创曲水风和
重庆
商业
Sunac Qushui Fenghe
Chongqing
Commercial

蓝城星翠里
浙江湖州
居住
Bluetown Xingcuili
Huzhou, Zhejiang
Residential

蓝城合山境
浙江湖州
居住
Bluetown Heshanjing
Huzhou, Zhejiang
Residential

合肥翡翠路图书馆景观设计
安徽合肥
景观设计
Hefei Feicui Road Library
Hefei, Anhui
Landscape Design

访溪上
浙江杭州
改造
Mr. Shang
Hangzhou, Zhejiang
Renovation

蓝城上海古北租赁房
上海
居住
Bluetown Shanghai Gubei
Rental Housing
Shanghai
Residential

浙商银行总部
浙江杭州
办公
China Zheshang Bank
Headquarters
Hangzhou, Zhejiang
Office

绿城凤起玉鸣
浙江温州
居住/教育
Greentown Phoenix Mansion
Wenzhou, Zhejiang
Residential/Education

融创哈尔滨融园
黑龙江哈尔滨
居住
Sunac Harbin Rongyuan
Harbin, Heilongjiang
Residential

绿城育华杨柳春风学校
山东济南
教育
Greentown Yuhua Willow Spring School
Jinan, Shandong
Education

舟山绿城育华幼儿园
浙江舟山
教育
Zhoushan Greentown Yuhua
Kindergarten
Zhoushan, Zhejiang
Education

毕玺文化吉安咖啡馆
江西吉安
商业
Bixi Culture Ji'an Cafe
Ji'an, Jiangxi
Commercial

中国西部军民融合创新谷暨西安电子谷
陕西西安
城市设计
The Western China Civil - Military
Integration Innovation Valley and
Xi'an Electronics Valley
Xi'an, Shanxi
Urban Design

滨江鹿城壹号
浙江温州
居住/教育
Binjiang Lucheng No.1
Wenzhou, Zhejiang
Residential/Education

绿城丽水绿谷信息产业园
浙江丽水
城市设计
Greentown Lishui Green Valley
Information Industry Park
Lishui, Zhejiang
Urban Design

蓝城金华浦江诗画小镇配套
浙江金华
文旅
Bluetown Jinhua Pujiang
Poetry and Painting Town
Amenities
Jinhua, Zhejiang
Cultural Tourism

台州鉴湖云舍
浙江台州
酒店
Taizhou Jianhu Yunshe
Taizhou, Zhejiang
Hospitality

绿城锦绣丽江·半山别墅
云南丽江
居住
Greentown Jinxiu Lijiang
Banshan Villa
Lijiang, Yunnan
Residential

蓝城宁波明庐
浙江余姚
居住
Bluetown Ningbo Ming Lu
Yuyao, Zhejiang
Residential

叶友吉安牛田村游客服务中心
江西吉安
商业
Yeyou Ji'an Niutian Village Tourist
Service Center
Ji'an, Jiangxi
Commercial

蓝城陶然里
浙江杭州
康养
Bluetown The Kidult
Hangzhou, Zhejiang
Health & Wellness

杭州华缘居
浙江杭州
居住
Hangzhou Hua Yuan Ju
Hangzhou, Zhejiang
Residential

北戴河悦榕庄酒店
河北秦皇岛
酒店
Banyan Tree Beidaihe
Qinhuangdao, Hebei
Hospitality

安吉龙山人文纪念园会所
浙江湖州
文化
Anji Longshan Humanism Memorial
Park Clubhouse
Huzhou, Zhejiang
Culture

蓝城红河东风韵小镇样板房
云南昆明
居住
Bluetown Red River Dongfengyun
Resettlement Housing
Kunming, Yunnan
Residential

融创延安宸院
陕西西安
居住
Sunac Yan'an Chenyuan
Xi'an, Shaanxi
Residential

融创苏州山水樾澜庭
江苏苏州
居住
Sunac Suzhou Landscape
Yuelan Court
Suzhou, Jiangsu
Residential

九龙仓杭州华发天荟
浙江杭州
居住
Wharf Hangzhou Huafa Tianhui
Hangzhou, Zhejiang
Residential

上海乡悦华亭田园综合体
上海
城乡协同
Shanghai Xiang Yue Hua Ting
Rural Complex
Shanghai
Suburban-urban Mutualism

融创苏州泊云庭
江苏苏州
居住
Sunac Suzhou Bravura
Suzhou, Jiangsu
Residential

广州融创文旅城璟府 B4 区
广东广州
居住
Sunac Guangzhou Land
Jingfu B4 District
Guangzhou, Guangdong
Residential

绿城南京小天堂站项目
江苏南京
居住
Greentown Nanjing Xiaotiantang
Station Residence
Nanjing, Jiangsu
Residential

仁恒海上源
上海
居住
Yanlord Arcadia
Shanghai
Residential

九龙仓上海杨浦辽源西路项目
上海
居住
Wharf Shanghai Yangpu Liaoyuan
West Road Project
Shanghai
Residential

绿城杭州花园岗街办公楼
浙江杭州
办公
Greentown Hangzhou Huayuangang
Street Office Building
Hangzhou, Zhejiang
Office

蓝城溪上锦棠里
浙江湖州
居住
Bluetown Valley Concertos
Huzhou, Zhejiang
Residential

天津观塘大厦
天津
办公
Tianjin Kwun Tong Building
Tianjin
Office

黄山旅游总部大厦
安徽黄山
办公
Huangshan Tourism
Headquarters Building
Huangshan, Anhui
Office

西双版纳悦景庄
云南西双版纳
居住
Xishuangbanna Yuejing Manor
Xishuangbanna, Yunnan
Residential

绿城宜春孝田住宅
江西宜春
居住
Greentown Yichun Xiaotian
Residence
Yichun, Jiangxi
Residential

融创郑州新村大道项目04地块
河南郑州
居住
Sunac Zhengzhou Xincun Avenue Plot
04 Residence
Zhengzhou, Henan
Residential

伟光汇通山东济宁任城区住宅
山东济宁
居住
Weiguang Huitong Shandong Jining
Rencheng District Residence
Jining, Shandong
Residential

杭州竺可桢中学
浙江杭州
教育
Hangzhou Chu Kochen
Honors School
Hangzhou, Zhejiang
Education

青岛海洋活力区|融创中心
山东青岛
居住
Qingdao Ocean Activities Zone |
Hopsca Center
Qingdao, Shandong
Residential

融创沈阳创新路住宅
辽宁沈阳
居住
Sunac Shenyang Chuangxin Road
Residence
Shenyang, Liaoning
Residential

西安瑞吉酒店概念设计
陕西西安
酒店
The St. Regis Xi'an
Xi'an, Shaanxi
Hospitality

天台山雪乐园
浙江台州
体育
Tiantaishan Snow Park
Taizhou, Zhejiang
Sports

陆家嘴锦绣御澜
上海
居住
Lujiazui Group Riverside Harmony
Shanghai
Residential

杭州西子智慧产业园东地块
浙江杭州
产业
Hangzhou Xizi Wisdom Industrial
Park East Plot
Hangzhou, Zhejiang
Industrial

绿城杭州萧山兴议村住宅
浙江杭州
居住
Greentown Hangzhou Xiaoshan Xingyi Village Residence
Hangzhou, Zhejiang
Residential

珠海横琴深蓝视界广场
广东珠海
办公/居住
Zhuhai Hengqin Deep Blue Square
Zhuhai, Guangdong
Office/Residential

融创重庆桃花源
重庆
居住
Sunac Chongqing Tao Hua Yuan
Chongqing
Residential

绿管湖北孝感金卉庄园住宅
湖北孝感
居住
Greentown Construction Management Xiaogan Jinhui Manor Residence
Xiaogan, Hubei
Residential

融创外滩壹号院
上海
居住
Sunac Bund One Sino Park
Shanghai
Residential

蓝城正大春风蓝湾家园
浙江宁波
居住
Bluetown & Zhengda Spring Breeze Blue Bay Home
Ningbo, Zhejiang
Residential

融创静安映
上海
居住
Sunac Jing'an Ying
Shanghai
Residential

海港新城唐山养老社区规划
河北唐山
规划
Haigang Xincheng Property Tangshan Pension Community Planning
Tangshan, Hebei
Planning

绿城川菜小镇
四川成都
文旅
Greentown Sichuan Cuisine Town
Chengdu, Sichuan
Cultural Tourism

绿城上海永丰街道住宅
上海
居住
Greentown Shanghai Yongfeng Street Residence
Shanghai
Residential

蓝城宿迁周马村新型社区
江苏宿迁
城乡协同
Bluetown Suqian Zhouma Village New Community
Suqian, Jiangsu
Suburban-urban Mutualism

北京朝阳公园·融创壹号院
北京
居住
One Sunac Opus on the Sun Park
Beijing
Residential

融创北京辛庄南路办公楼
北京
办公
Sunac Beijing South Xinzhuang Road Office
Beijing
Office

绿城南通晓风印月
江苏南通
居住
Nantong Greentown Collection
Nantong, Jiangsu
Residential

2019

绿管中证南京奶山项目
江苏南京
规划
Greentown Construction Management CSI Nanjing Naishan Mountain Project
Nanjing, Jiangsu
Planning

绿城重庆晓风印月
重庆市
居住
Chongqing Greentown Collection
Chongqing
Residential

绿城潍坊桃李春风
山东潍坊
居住/教育
Greentown Weifang The Spring Blossom
Weifang, Shandong
Residential/Education

蓝城汉中春风江南国际颐养社区
陕西汉中
居住
Bluetown Hanzhong Chunfeng Jiangnan international Nursing Community
Hanzhong, Shaanxi
Residential

融创滁州乌衣古镇
安徽滁州
文旅
Sunac Chuzhou Wuyi Ancient Town
Chuzhou, Anhui
Cultural Tourism

万星千岛湖天清岛度假别墅改造
浙江杭州
改造
Wanxing Thousand Island Lake Tianqing Island Resort Villa Renovation
Qiandao Lake, Zhejiang
Renovation

绿城上海崇明城桥镇住宅
上海
居住
Greentown Shanghai Chongming Chengqiao Town Residence
Shanghai
Residential

融创杭州森与海之城
浙江杭州
文旅
Sunac Hangzhou Sayhi
Hangzhou, Zhejiang
Cultural Tourism

融创·历城控股济南东山府
山东济南
居住
Sunac & Licheng Holding Jinan Eastward Calm Mansion
Jinan, Shandong
Residential

城投华润天津瑞府
天津
居住
Chengtou & CR Land Tianjin Mansion
Tianjin
Residential

融创江阴芙蓉大道住宅
江苏无锡
居住
Sunac Jiangyin Furong Avenue Residence
Wuxi, Jiangsu
Residential

中铁交投花语天境
浙江杭州
居住
China Railway Communications Investment Glory Mansion
Hangzhou, Zhejiang
Residential

滨耀城
浙江杭州
混合开发
Colorful City
Hangzhou, Zhejiang
Mixed-use

绿城大连玫瑰园葡萄酒小镇
辽宁大连
文旅
Greentown Dalian Cheers Valley
Dalian, Liaoning
Cultural Tourism

融创滨江杭源御潮府
浙江杭州
居住
Sunac Binjiang Imperial Mansion
Hangzhou, Zhejiang
Residential

绿城宁波凤起潮鸣
浙江宁波
居住/商业
Greentown Ningbo Phoenix Mansion
Ningbo, Zhejiang
Residential/Commercial

绿城南京栖霞区北苑东路住宅
江苏南京
居住
Greentown Nanjing Qixia District East Beiyuan Road Residence
Nanjing, Jiangsu
Residential

绿城哈尔滨杨柳郡
黑龙江哈尔滨
居住
Greentown Harbin Young Ctiy
Harbin, Heilongjiang
Residential

融创天津泽川路项目
天津
居住/商业
Sunac Tianjin Zechuan Road Project
Tianjin
Residential/Commercial

287

融创长乐雅颂
重庆
居住
Sunac Changle Yasong
Chongqing
Residential

融创重庆中航小镇
重庆
文旅
Sunac Chongqing AVIC Town
Chongqing
Cultural Tourism

融创上海十里江湾
上海
居住
Sunac Shanghai River Villa
Shanghai
Residential

绿城宁波吉瑞府
浙江宁波
居住
Greentown Ningbo Jirui Mansion
Ningbo, Zhejiang
Residential

华润温州九悦
浙江温州
居住
China Resources Wenzhou Jiuyue
Wenzhou, Zhejiang
Residential

绿城宁波吉好商业
浙江宁波
商业
Greentown Ningbo Jihao Commerce
Ningbo, Zhejiang
Commercial

中国铁建珠海国际城
广东珠海
居住
CRCC International City
Zhuhai, Guangdong
Residential

祥符桥传统风貌街区
浙江杭州
城市更新
Xiangfu Bridge Historic District Renewal
Hangzhou, Zhejiang
Urban Renewal

张謇故里小镇柳西半街
江苏南通
城乡协同
Jianli Town Liuxiban Street
Nantong, Jiangsu
Suburban-urban Mutualism

融创银川永泰城住宅
宁夏银川
居住
Sunac Yinchuan Yongtai City Residence
Yinchuan, Ningxia
Residential

同盾杭州未来科技城办公楼
浙江杭州
办公
Tongdun Hangzhou Future Sci-Tech City Office Building
Hangzhou, Zhejiang
Office

融创内蒙古赛罕区西把栅项目
内蒙古呼和浩特
居住
Sunac Inner Mongolia Saihan District Xibazha Residence
Hohhot, Inner Mongolia
Residential

融创南京迈尧西路住宅
江苏南京
居住
Sunac Nanjing West Maiyao Road Residence
Nanjing, Jiangsu
Residential

金华双龙宾馆改造
浙江金华
改造/酒店
Jinhua Shuanglong Hotel Renovation
Jinhua, Zhejiang
Renovation/Hospitality

贵阳英迪格酒店
贵州贵阳
酒店
Hotel Indigo Guiyang
Guiyang, Guizhou
Hospitality

中南西安凤城四路住宅
陕西西安
居住
Zhongnan Group Xi'an Fengcheng No.4 Road Residence
Xi'an, Shaanxi
Residential

蓝城杭州联乐名苑
浙江杭州
居住
Bluetown Hangzhou Lianle Mingyuan
Hangzhou, Zhejiang
Residential

融创西安海逸长洲
陕西西安
居住
Sunac Xi'an Great Center Park
Xi'an, Shaanxi
Residential

蓝城杭州山秀景苑
浙江杭州
居住
Bluetown Hangzhou Shanxiu Jingyuan
Hangzhou, Zhejiang
Residential

蓝城杭州滨江长河社区安置房
浙江杭州
居住
Bluetown Hangzhou Binjiang Changhe Community Resettlement Housing
Hangzhou, Zhejiang
Residential

耀达临海松溪住宅
浙江临海
居住
Yaoda Seaward Songxi Residence
Linhai, Zhejiang
Residential

绿城外滩兰庭
上海
居住
Greentown The Bund Garden
Shanghai
Residential

融创太仓娄江新城官河路住宅
江苏太仓
居住
Sunac Taicang Loujiang New Town Guanhe Road Residence
Taicang, Jiangsu
Residential

物产中大德清明星村酒店
浙江湖州
酒店
WZ Group Deqing Star Village Hotel
Huzhou, Zhejiang
Hospitality

蓝城春风江山
浙江江山
居住
Bluetown Poetic Jiangshan
Jiangshan, Zhejiang
Residential

温哥华Lot10地块会所
加拿大温哥华
酒店
Vancouver Plot10 Club
Vancouver, Canada
Hospitality

绿城大连东北路住宅
辽宁大连
居住
Greentown Dalian Northeast Road Residence
Dalian, Liaoning
Residential

绿城杭州未来科技城良上路住宅
浙江杭州
居住
Greentown Hangzhou Liangshang Road Future Sci-Tech City Residence
Hangzhou, Zhejiang
Residential

绿城春风金沙
浙江杭州
居住
Greentown Hangzhou Lakeside Mansion
Hangzhou, Zhejiang
Residential

绿城无锡诚园
江苏无锡
居住
Greentown Wuxi Sincere Garden
Wuxi, Jiangsu
Residential

融创重庆朝天门项目
重庆
混合开发
Sunac Chongqing Chaotianmen Residence
Chongqing
Mixed-use

石家庄中央商务区办公楼
河北石家庄
办公
Shijiazhuang CBD Office Towers
Shijiazhuang, Hebei
Office

昆明贵宾楼改造
云南昆明
改造
Kunming Guest Builidng Renovation
Kunming, Yunnan
Renovation

石家庄融创时代中心
河北石家庄
居住
Shijiazhuang Sunac Times Center
Shijiazhuang, Hebei
Residential

保利华侨城云禧
广东佛山
居住
Poly Chief Palace
Foshan, Guangdong
Residential

三峡游轮中心
湖北宜昌
混合开发
The Three Gorges Cruise Center
Yichang, Hubei
Mixed-use

绿管山东耕读旅居小镇规划
山东曲阜
规划
Greentown Construction Management
Shandong Gengdu Holiday Town
Planning
Qufu, Shandong
Planning

贵阳中天·吾乡
贵州贵阳
居住
Guiyang Zhongtian Wuxiang
Guiyang, Guizhou
Residential

融创哈尔滨通郊南路住宅
黑龙江哈尔滨
居住
Sunac Harbin South Tongjiao
Road Residence
Harbin, Heilongjiang
Residential

绿城美的广州晓风印月
广东广州
居住
Greentown & Midea Guangzhou
Greentown Collection
Guangzhou, Guangdong
Residential

融创河北唐山火炬路住宅
河北唐山
居住
Sunac Hebei Tangshan Huoju
Road Residence
Tangshan, Hebei
Residential

中企云萃江湾
上海
居住
China Enterprise International
Yuncui Bay
Shanghai
Residential

鹏欣昆明呈贡雨花区住宅
云南昆明
居住
Pengxin Kunming Chenggong
Yuhua District Residence
Kunming, Yunnan
Residential

信创电梯产业园
浙江杭州
产业/办公
Xinchuang Elevator Industrial Park
Hangzhou, Zhejiang
Industrial/Office

融创郑州嘉和小镇 B-04/07/09/16/18 地块
河南郑州
居住
Sunac Zhengzhou Jiahe Town
B-04/07/09/16/18 Plot Residence
Zhengzhou, Henan
Residential

华润上海周泾村项目
上海
规划
China Resources Shanghai
Zhoujing Village Residence
Shanghai
Planning

绿城福州闽侯青口项目
福建福州
居住
Greentown Fuzhou Minhou
Qingkou Residence
Fuzhou, Fujian
Residential

融创云南抚仙湖国际度假小镇
云南玉溪
文旅
Sunac Yunnan Fuxian Lake
International Resort
Yuxi, Yunnan
Cultural Tourism

杭州嘉里城
浙江杭州
混合开发
Hangzhou Kerry Parkside
Hangzhou, Zhejiang
Mixed-use

融创上海赵巷业煌路南侧住宅
上海
居住
Sunac Shanghai Zhaoxiang South
Yehuang Road Residence
Shanghai
Residential

中交重庆翠澜羣境
重庆
居住
CCCG Chongqing Grand Space
Chongqing
Residential

昆仑哈尔滨崂山路项目
黑龙江哈尔滨
居住
Kunlun Harbin Laoshan Road Residence
Harbin, Heilongjiang
Residential

绿城南京仙林湖住宅
江苏南京
居住
Greentown Nanjing Xianlin Lake
Residence
Nanjing, Jiangsu
Residential

融创沈阳十五号街住宅
辽宁沈阳
居住
Sunac Shenyang No. 15 Street Residence
Shenyang, Liaoning
Residential

绿城晓月澄庐
浙江杭州
居住
Greentown Moon and House
Hangzhou, Zhejiang
Residential

望江中心TOD
浙江杭州
TOD
Wangjiang Center TOD
Hangzhou, Zhejiang
TOD

蓝都桐乡凤凰湖酒店
浙江桐乡
酒店
Landu Tongxiang Phoenix Lake Hotel
Tongxiang, Zhejiang
Hospitality

信创杭州泉口街厂房
浙江杭州
产业
Xinchuang Hangzhou Quankou
Street Factory
Hangzhou, Zhejiang
Industrial

中交昆明锦澜府
云南昆明
居住
CCCG Kunming Splendid Mansion
Kunming, Yunnan
Residential

石家庄汀香郡
河北石家庄
居住/教育
Shijiazhuang Riverside Town
Shijiazhuang, Hebei
Residential/Education

绿城徐州孟庄住宅
江苏徐州
居住
Greentown Xuzhou Mengzhuang
Residence
Xuzhou, Jiangsu
Residential

兰溪服务区
浙江兰溪
交通
Lanxi Expressway Service Area
Lanxi, Zhejiang
Transpotation

曹山未来城古桥水镇
江苏常州
文旅
Caoshan Future City Guqiao Water Town
Changzhou, Jiangsu
Cultural Tourism

曹山未来城古桥水镇亲子活动中心
江苏常州
文化
Caoshan Future City Guqiao
Watertown Family Center
Changzhou, Jiangsu
Culture

蓝城滨河湾
安徽合肥
居住
Bluetown Mansion Luxury
Hefei, Anhui
Residential

绿城西安陈林村居住 19/22 地块
陕西西安
居住
Greentown Xi'an Chenlin Village
Plot 19/22 Residence
Xi'an, Shaanxi
Residential

绿城石家庄桂语听澜
河北石家庄
居住
Greentown Shijiazhuang
Guiyu Tinglan
Shijiazhuang, Hebei
Residential

融创咸阳御河宸院
陕西咸阳
居住
Sunac Xianyang River Courtyards
Xianyang, Shaanxi
Residential

蓝城昆明抚仙湖项目
云南昆明
居住
Bluetown Kunming Fuxian Lake Residence
Kunming, Yunnan
Residential

绿城宁波春熙月鸣
浙江宁波
居住
Greentown Ningbo Spring Moon
Ningbo, Zhejiang
Residentia

绿管重庆太龙镇项目
重庆
居住
Greentown Construction Management Chongqing Tailong Town Residence
Chongqing
Residential

绿城西安幸福林带安置房
陕西西安
居住 / 教育
Greentown Xi'an XingFu Green Belt Settlement Housing
Xi'an, Shaanxi
Residential/Education

中交舟山南山美庐立面设计
浙江舟山
居住
CCCG Zhoushan Nanshan Beautiful Cottage Facade Design Zhoushan, Zhejiang
Residential

圩美磨滩精品酒店
安徽合肥
酒店
Weimei Coast Boutique Hotel
Hefei, Anhui
Hospitality

新东城太原东景庭院
山西太原
居住
New East City Taiyuan East View Courtyard
Taiyuan, Shanxi
Residential

绿城春来枫华
浙江杭州
居住
Greentown Chunlai Fenghua
Hangzhou, Zhejiang
Residential

里直街
浙江绍兴
城市更新
Lizhi Street
Shaoxing, Zhejiang
Urban Renewal

雅达阳羡溪山三期
江苏无锡
居住
Yangxian Landscape Phase III
Wuxi, Jiangsu
Residential

雅达医院
江苏无锡
医疗
Yada Hospital
Wuxi, Jiangsu
Healthcare

蓝城义乌大陈小镇中心
浙江义乌
规划 / 商业
Bluetown Yiwu Dacheng Village Center
Yiwu, Zhejiang
Planning/Commercial

上海塘沽路城市更新
上海
城市更新 / 规划
Shanghai Tanggu Road Renewal
Shanghai
Urban Renewal/Planning

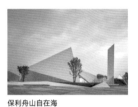

保利舟山自在海
浙江舟山
文旅
Poly Zhoushan Freedom Sea
Zhoushan, Zhejiang
Cultural Tourism

金桥金鼎·阅府
上海
居住
Golden Bridge Master Cube
Shanghai
Residential

雅达剧院
江苏宜兴
文化
Yada Theater
Yixing, Jiangsu
Culture

绿城杭州余杭区双联村项目
浙江杭州
规划
Greentown Hangzhou Yuhang District Shuanglian Village Residence
Hangzhou, Zhejiang
Planning

济南融创文旅城D18地块
山东济南
居住
Jinan Sunac Land Plot 18 Residence
Jinan, Shandong
Residential

绿城南通江纬路住宅
江苏南京
居住
Greentown Nantong Jiangwei Road Residence
Nanjing, Jiangsu
Residential

绿城郑州湖畔云庐
河南郑州
居住
Greentown Zhengzhou Oriental Villa
Zhengzhou, Henan
Residential

融创上海临港科学家社区
上海
规划
Sunac Shanghai Lingang Scientists Community
Shanghai
Planning

西湖天地杭州南山路项目
浙江杭州
酒店
West Lake Mall Hangzhou Nanshan Road Project
Hangzhou, Zhejiang
Hospitality

蓝城湖州春和云树里
浙江湖州
居住
Bluetown Huzhou Chuihe Yunshuli
Huzhou, Zhejiang
Residential

中交重庆翠澜逸境
重庆
居住
CCCG Chongqing Luxury Space
Chongqing
Residential

众安杭州金沙湖项目
浙江杭州
居住
Zhong'an Hangzhou Jinsha Lake Residence
Hangzhou, Zhejiang
Residential

明月松间·潜川
浙江杭州
酒店
Yakamoz Qianchuan
Hangzhou, Zhejiang
Hospitality

融创哈尔滨文景壹号
黑龙江哈尔滨
居住
Sunac Harbin Wenjing No.1
Harbin, Heilongjiang
Residential

中建尼山圣地鲁源小镇
山东济宁
文旅
China State Construction Nishan Holy Land Luyuan Town
Jining, Shandong
Cultural Tourism

西子钱塘智慧产业园
浙江杭州
产业 / 办公
Xizi Qiantang Intelligence Industrial Park
Hangzhou, Zhejiang
Industrial/Office

融创青岛崂山壹号院
山东青岛
居住
Sunac Qingdao Laoshan No. 1 Yard
Qingdao, Shandong
Residential

蓝城汉中春风江南
陕西汉中
居住 / 商业 / 教育
Bluetown Hanzhong Spring Breeze
Hanzhong, Shaanxi
Residential/Commercial/Education

绿城江河鸣翠
浙江杭州
居住
Greentown Hangzhou River Collection
Hangzhou, Zhejiang
Residential

2020

杭州航海金座
浙江杭州
办公
Hangzhou Voyage Jinzuo
Hangzhou, Zhejiang
Office

众安温州顺源里
浙江温州
居住
Zhong'an Wenzhou Shunyuan Li
Wenzhou, Zhejiang
Residential

绿城大连竹青街住宅
辽宁大连
居住
Greentown Dalian Zhuqing Street Residence
Dalian, Liaoning
Residential

中交重庆蔚蓝天镜
重庆
居住
CCCG Chongqing Weilan Tianjing
Chongqing
Residential

融创北京香山壹号院
北京
居住
Sunac Beijing One Sino Park
Beijing
Residential

杭锅产业园西地块
浙江杭州
产业/办公
Hangzhou Boiler
Industrial Park West Lot
Hangzhou, Zhejiang
Industrial/Office

蓝城义乌桃李春风
浙江义乌
文旅
Bluetown Yiwu The Spring Blossom
Yiwu, Zhejiang
Cultural Tourism

绿城北京奥海明月
北京
居住
Greentown Beijing Bright
Moon in Forest
Beijing
Residential

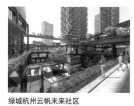

绿城杭州云帆未来社区
浙江杭州
城市设计
Greentown Hangzhou Yunfan
Future Community
Hangzhou, Zhejiang
Urban Design

绿城北京明月听兰
北京
居住
Greentown Beijing Orchid Garden
Beijing
Residential

恒力环企中心
江苏苏州
混合开发
Hengli Global Enterprise Center
Suzhou, Jiangsu
Mixed-use

华润北京海淀幸福里
北京
居住
CR Land Beijing Park Lane Mansion
Beijing
Residential

中交武汉泓园
湖北武汉
居住
CCCG Wuhan Luxury Aesthetic
Wuhan, Hubei
Residential

绿城无锡宸风云庐
江苏无锡
居住
Greentown Wuxi Trees Villa
Wuxi, Jiangsu
Residential

万科上海马桥镇住宅
上海
居住
Vanke Shanghai Maqiao Town Residence
Shanghai
Residential

广州御溪臻山墅
广东广州
居住
Guangzhou Villa of Moutain
Guangzhou, Guangdong
Residential

蓝城上虞运河江南里二期
浙江绍兴
文旅
Bluetown Shangyu Jiangnanli Phase II
Shaoxing, Zhejiang
Cultural Tourism

蓝城上海嘉定联华村路住宅
上海
居住
Bluetown Shanghai Jiading
Lianhuacun Road Residence
Shanghai
Residential

绿城秦皇岛戴河大街住宅
河北秦皇岛
居住
Greentown Qinhuangdao Daihe
Avenue Residence
Qinhuangdao, Hebei
Residential

杭州亚运村社区公园
浙江杭州
文化
Hangzhou Asian Sports Village
Communicty Park
Hangzhou, Zhejiang
Culture

绿城金华多湖街道住宅
浙江金华
居住
Greentown Jinhua Duohu Streen
Jinhua, Zhejiang
Residential

蓝城上虞丰惠镇项目
浙江绍兴
居住
Bluetown Shangyu Fenghui
Town Residence
Shaoxing, Zhejiang
Residential

中交北京门头沟住宅
北京
居住
CCCG Beijing Mentougou Residence
Beijing
Residential

融通北京东大桥路办公楼
北京
办公
CRTC Beijing Dongdaqiao Road
Office Building
Beijing
Office

西子专精特新产业园
浙江杭州
产业/办公
Xizi Specialized and Innovative
Industrial Park
Hangzhou, Zhejiang
Industrial/Office

上海西岸金融城G地块
上海
混合开发/城市更新
Shanghai West Bund Financial
City - Plot G
Shanghai
Mixed-use/Urban Renewal

杭州显宁寺
浙江杭州
文化
Hangzhou Xianning Temple
Hangzhou, Zhejiang
Culture

仁恒马鞍山和县香泉湖文旅项目
安徽马鞍山
规划
Yanlord Ma'anshan Hexian Xiangquan
Lake Cultural Tourism Project
Ma'anshan, Anhui
Planning

上海九棵树未来艺术中心改造
上海
文化
Renovation of Shanghai Jiukeshu
Art Center
Shanghai
Culture

阿里巴巴杭州爱橙路办公楼
浙江杭州
办公
Alibaba Hangzhou Aicheng Road Office
Building
Hangzhou, Zhejiang
Office

绿城乌鲁木齐明月兰庭
新疆乌鲁木齐
居住
Greentown Urumqi Orchid Yard
Urumqi, Xinjiang
Residential

上海乔家路项目
上海
城市更新
Shanghai Qiaojia Road Project
Shanghai
Urban Renewal

阿朵小镇2号酒店
山东青岛
酒店
A Dream A Life No.2 Hotel
Qingdao, Shandong
Hospitality

绿城天津桃李春风
天津
居住
Greentown Tianjin The
Spring Blossom
Tianjin
Residential

绿城昆明西南林业大学地块
云南昆明
居住
Greentown Kunming Southwest Forestry
University Plot Residence
Kunming, Yunnan
Residential

信创九江跨境电子商务产业园立面改造
江西九江
改造
Xinchuang Jiujiang Cross-border
E-commerce Industrial Park Facade
Renovation
Jiujiang, Jiangxi
Renovation

临港上海金融湾
上海
居住
Lingang Shanghai Financial Bay
Shanghai
Residential

蓝城上海嘉兴公路住宅
上海
居住
Bluetown Shanghai Jiaxin Highway Residence
Shanghai
Residential

杭州云城双铁上盖区域综合开发
浙江杭州
TOD
Transit Oriented Development Architectural Concepts for the Superstructure of Hangzhou Yuncheng
Hangzhou, Zhejiang
TOD

绿城绍兴鉴湖路住宅
浙江绍兴
居住
Greentown Shaoxing Jianhu Road Residence
Shaoxing, Zhejiang
Residential

蓝城德清青马住宅
浙江湖州
规划
Bluetown Deqing Qingma Residence
Huzhou, Zhejiang
Planning

星野成都办公园区
四川成都
办公
Xingye Chengdu Taihe Road Office Park
Chengdu, Sichuan
Office

绿城空中院墅样板间
浙江杭州
居住
Greentown Sky Villa Model Unit
Hangzhou, Zhejiang
Residential

绿城武汉湖畔云庐
湖北武汉
混合开发
Greentown Wuhan Oriental Villa
Wuhan, Hubei
Mixed-use

杭锅崇贤分公司停车楼扩建
浙江杭州
产业
Hangzhou Boiler Group Chongxian Branch Parking Building Expansion
Hangzhou, Zhejiang
Industrial

融创西安时代奥城
陕西西安
居住
Sunac Xi'an Times City
Xi'an, Shaanxi
Residential

璟盛丽江金虹路项目
云南丽江
规划
Jingsheng Lijiang Jinhong Road Project
Lijiang, Yunnan
Planning

绿城吉祥里雅泸名筑
浙江杭州
居住
Greentown Joy in Block Yalu Mingzhu
Hangzhou, Zhejiang
Residential

宁波智造港芯创园
浙江宁波
产业/办公
Ningbo Intelligent Manufacturing Port
Ningbo, Zhejiang
Industrial/Office

绿城宁波中兴大桥文创港
浙江宁波
规划
Greentown Ningbo Zhongxing Bridge Cultural Innovation Harbor
Ningbo, Zhejiang
Planning

杭州市西湖区龙坞茶镇核心区规划
浙江杭州
规划
Hangzhou Xihu District Longwucha Tow Core Area Planing
Hangzhou, Zhejiang
Planning

华夏幸福深圳康复医疗中心
广东深圳
医疗
CFLD Shenzhen Medical Rehabilitation Center
Shenzhen, Guangdong
Healthcare

绿管南平金融产业小镇
福建南平
规划
Greentown Construction Management Nanping Financial Industrial Town
Nanping, Fujian
Planning

绿城宁波春来晓园
浙江宁波
居住
Greentown Ningbo Oriental Dawn
Ningbo, Zhejiang
Residential

交投杭州钱江世纪城金融总部大楼
浙江杭州
办公/规划
CICO Hangzhou Qianjiang Century City Finance Headquarters
Hangzhou, Zhejiang
Office/Planning

义乌建投秋实路安置房
浙江金华
居住
Yiwu Construction Investment Settlement Yiwu Qiushi Road Resettlement Housing
Jinhua, Zhejiang
Residential

绿城廊坊下庄头一号路住宅
河北廊坊
居住
Greentown Langfang Xiazhuang No.1 Road Residence
Langfang, Hebei
Residential

滨江杭州海潮望月城
浙江杭州
居住/办公
Binjiang Hangzhou Luna Mansion
Hangzhou, Zhejiang
Residential/Office

绿管富阳银湖街道安置房
浙江杭州
居住
Greentown Construction Management Fuyang Yinhu Road Resettlement Housing
Hangzhou, Zhejiang
Residential

众安杭州九和路项目
浙江杭州
商业
Zhong'an Hangzhou Jiuhe Road Project
Hangzhou, Zhejiang
Commercial

融创上海五坊园项目
城市更新
上海
Sunac Shanghai Fiverows Garden Project
Urban Renewal
Shanghai

绿地上海金陵东路项目
上海
居住
Greenland Shanghai East Jinling Road Residence
Shanghai
Residential

交投丽水栖溪晓庐
浙江丽水
居住
CICO Majestic Mansion
Lishui, Zhejiang
Residential

蓝城上海青浦涵璧湾二期
上海
规划
Bluetown Shanghai Qingpu District Hanbi Bay Residence Phase II
Shanghai
Planning

上海漕河泾人工智能小镇
上海
文旅/产业
Shanghai Caohejing Artificial Intelligence Town
Shanghai
Cultural Tourism/Industrial

绿城重庆照母山养老项目
重庆
规划
Greentown Chongqing Zhaomu Mountain Healthcare Center
Chongqing
Planning

上海北外滩32街坊更新
上海
城市更新
Shanghai North Bund Neighborhood 32 Renewal
Shanghai
Urban Renewal

阿朵小镇美术馆
山东青岛
文化
Sunac A Dream A Life Art Museum
Qingdao, Shandong
Culture

杭锅德清伟业路厂房
浙江杭州
产业/办公
Hangzhou Boiler Plant Deqing Weiye Road Factory
Hangzhou, Zhejiang
Industrial/Office

信创杭州泉口街二期厂房
浙江杭州
产业
Xinchuang Hangzhou Quankou Street Plant Phase II
Hangzhou, Zhejiang
Industrial

融创咸阳时光宸阆
陕西咸阳
居住
Sunac Xianyang Power of Center
Xianyang, Shaanxi
Industrial

中交长安里
陕西咸阳
居住
CCCG Chang'an Mansion
Xianyang, Shaanxi
Residential

沪杭高速嘉兴服务区
浙江嘉兴
交通
G60 Expressway Jiaxing Service Area
Jiaxing, Zhejiang
Transpotation

苏州狮山悦榕庄
江苏苏州
酒店
Banyan Tree Suzhou Shishan
Suzhou, Jiangsu
Hospitality

老板电器杭州总部办公楼
浙江杭州
办公
ROBAM Hangzhou Headquarters
Hangzhou, Zhejiang
Office

融创重庆南滨路融创湾项目
重庆
混合开发
Sunac Chongqing Nanbin Road Sunac Bay
Chongqing
Mixed-use

阿里巴巴杭州西溪董湾项目
浙江杭州
酒店
Alibaba Hangzhou Dongwan Xixi Hotel
Hangzhou, Zhejiang
Hospitality

银润蓝城·天使小镇
浙江湖州
居住
Yinrun Lancheng Town of Angels
Huzhou, Zhejiang
Residential

滨江桐庐富春未来城洋洲南路项目
浙江杭州
办公/居住
Binjiang Tonglu Fuchun Future City South Yangzhou Road Project
Hangzhou, Zhejiang
Office/Residential

融创哈尔滨乐园街项目
黑龙江哈尔滨
居住
Sunac Harbin Leyuan Street Residence
Harbin, Heilongjiang
Residential

蓝城杭州半岛花园
浙江杭州
居住
Bluetown Hangzhou Peninsula Garden
Hangzhou, Zhejiang
Residential

中海顺昌玖里
上海
城市更新/居住/商业
China Overseas Arbour
Shanghai
Urban Renewal/Residential/Commercial

绿城新湖启东圆陀角项目
江苏南通
居住/商业
Greentown Xinhu Qidong Yuantuojiao Residence Nantong, Jiangsu
Residential/Commercial

绿城成都三圣乡白桦林路住宅
四川成都
居住
Greentown Chengdu Sansheng County Baihualin Road Residence
Chengdu, Sichuan
Residential

蓝城杭州长乐农庄项目
浙江杭州
居住
Bluetown Hangzhou Changle Farm Residence
Hangzhou, Zhejiang
Residential

绿城成都青白江云溪漫谷项目
四川成都
居住
Greentown Chengdu Qingbai River Yunxi Diffuse Valley Planning
Chengdu, Sichuan
Residential

海口五源河创新产业中心
海南海口
产业/办公
Haikou Wuyuanhe Intelligent Creative Collective
Haikou, Hainan
Industrial/Office

绿城乌鲁木齐中豪·润园
新疆乌鲁木齐
居住
Greentown Urumqi Zhong Hao Run Garden
Urumqi, Xinjiang
Residential

雅达阳羡溪山四期
江苏无锡
居住
Yada Yangxian Landscape Phase IV
Wuxi, Jiangsu
Residential

义乌建投雪峰西路安置房
浙江金华
居住
Yiwu Construction Investment West Xuefeng Road Settlement Building
Jinhua, Zhejiang
Residential

华邦扬州珑玥湖小镇
江苏扬州
文旅
Huabang Yangzhou Longyue Lake town
Yangzhou, Jiangsu
Cultural Tourism

融创上海徐汇区黄石路地块
上海
城市更新/规划
Sunac Shanghai Xuhui District Huangshi Road Urban Renewal
Shanghai
Urban Renewal/Planning

华润宁波江望悦府
浙江宁波
混合开发
China Resources Ningbo Jiangwangyue Mansion
Ningbo, Zhejiang
Mixed-use

紫薇绿城西安南山云庐
陕西西安
居住
Ziwei & Greentown Xi'an Hills Villa
Xi'an, Shaanxi
Residential

绿城佛山桂语兰庭
广东佛山
居住
Greentown Foshan Guiyu Lan Garden
Foshan, Guangdong
Residential

绿城衢州高铁新城鹿鸣社区
浙江衢州
城市设计
Greentown Quzhou Luming Community High-speed Railway City
Quzhou, Zhejiang
Urban Design

蓝城上海花海芳香小镇
上海
文旅
Bluetown Shanghai Aroma Town
Shanghai
Cultural Tourism

杭州西湖高尔夫私家别墅
浙江杭州
居住
Hangzhou West Lake Golf Private Villa
Hangzhou, Zhejiang
Residential

德清莫干山洲际酒店
浙江湖州
酒店
InterContinental Deqing Moganshan
Huzhou, Zhejiang
Hospitality

中交武汉沿江大道项目
湖北武汉
规划
CCCG Wuhan Yanjiang Avenue Project
Wuhan, Hubei
Planning

曲水善湾乡村振兴示范区
江苏苏州
城乡协同
Qushui Shanwan Rural Revitalization Demonstration Area
Suzhou, Jiangsu
Suburban-urban Mutualism

华润武汉瑞府
湖北武汉
居住
CR Land Wuhan Park Lane Mansion
Wuhan, Hubei
Residential

仁恒上海晋元路项目
上海
居住
Yanlord Shanghai Jinyuan Road Residence
Shanghai
Residential

华德福平湖新仓中学
浙江平湖
教育
Huade Fupinghu Xincang Middle School
Pinghu, Zhejiang
Education

融创咸阳西部湾
陕西咸阳
居住
Sunac Xianyang Western Bay
Xianyang, Shaanxi
Residential

蓝城绿城鹰潭鹤鸣溪谷
江西鹰潭
居住
Bluetown & Greentown Yingtan Heming Valley
Yingtan, Jiangxi
Residential

华滋上海罗甸镇潘泾路项目
上海
规划
Huazi Shanghai Luodian Town
Panjing Road Project
Shanghai
Planning

绿城杭州湖上春风里
浙江杭州
居住
Greentown Hangzhou Hushang
Chunfengli
Hangzhou, Zhejiang
Residential

融创绍兴黄酒小镇
浙江绍兴
文旅
Sunac Shaoxing Wine Town
Shaoxing, Zhejiang
Cultural Tourism

海南五指山酒店
海南五指山
酒店
Hainan Wuzhi Mountain Hotel
Wuzhishan, Hainan
Hospitality

绿城复地杭州·月映咏荷园
浙江杭州
居住
Forte & Greentown Hangzhou Yueying
Yonghe Garden
Hangzhou, Zhejiang
Residential

奥正临沂琅琊园
山东临沂
规划/居住/酒店
Auzen Linyi Langya Park
Linyi, Shandong
Planning/Residential/Hospitality

绿城苏州吴江科技创业园
江苏苏州
规划
Greentown Suzhou Wujiang Science
and Technology Pioneer Park
Suzhou, Jiangsu
Planning

蓝城郑州桃花源
河南郑州
居住
Bluetown Zhengzhou Taohuayuan
Zhengzhou, Henan
Residential

富士康上海华漕国际办公楼
上海
办公
Foxconn Shanghai Huacao
Internatioanl Office Building
Shanghai
Office

巨基石家庄石钢西厂项目
河北石家庄
规划
Juji Shijiazhuang Shigang West
Plant Residence
Shijiazhuang, Hebei
Planning

绿城杭州天地墅园首府
浙江杭州
居住
Greentown Hangzhou Tiandi Villa Garden
Capital
Hangzhou, Zhejiang
Residential

2021

重庆启元
重庆
混合开发
Century Land Chongqing
Chongqing
Mixed-use

绿城杭州桂语山澜轩
浙江杭州
居住
Greentown Hangzhou Guiyu
Mountain Lanxuan
Hangzhou, Zhejiang
Residential

绿城西安和庐
陕西西安
居住
Greentown Xi'an Classic Villa
Xi'an, Shanxi
Residential

滨江杭州钱潮鸣翠云筑
浙江杭州
居住
Binjiang Hangzhou Time Melody
Hangzhou, Zhejiang
Residential

中新南京生态科技岛学校
江苏南京
教育
Zhongxin Nanjing Ecological
Technology Island School
Nanjing, Jiangsu
Education

华发长安首府
陕西西安
居住
Huafa Legend Chief
Xi'an, Shaanxi
Residential

大华金河湾9号楼
江西萍乡
居住
Dahua Jin He Wan No.9 Building
Pingxiang, Jiangxi
Residential

华润杭桢未来中心
浙江杭州
居住
CR Land Future Center
Hangzhou, Zhejiang
Residential

深圳万创云汇01-03地块
广东深圳
混合开发
Shenzhen Vanke Cloud Gradus
Plot 01-03
Shenzhen, Guangdong
Mixed-use

湖州莫干山高新区美憬阁酒店
浙江湖州
酒店
Huzhou Moganshan High Tech Zone
MGallery
Huzhou, Zhejiang
Hospitality

融创济南财富花园B1地块
山东济南
居住
Sunac Jinan Fortune Garden Lot B1 Project
Jinan, Shandong
Residential

绿城宁波鄞奉路住宅
浙江宁波
居住
Greentown Ningbo Yinfeng
Road Residence
Ningbo, Zhejiang
Residential

蓝城黄山歙县欧洲之星康养小镇
安徽黄山
康养
Bluetown Huangshan She County
Eurostar Wellness Town
Huangshan, Anhui
Health & Wellness

绿城大连旅顺北路项目
辽宁大连
规划
Greentown Dalian North Lyushun
Road Proejct
Dalian, Liaoning
Planning

瑞安上海西藏南路项目
上海
城市更新
Shui On Shanghai South Xizang
Road Project
Shanghai
Urban Renewal

湖州莫干山高新区美憬阁酒店
浙江湖州
酒店
Huzhou Moganshan High Tech
Zone MGallery
Huzhou, Zhejiang
Hospitality

义乌双江湖江东街道毛店安置房
浙江金华
居住
Yiwu Shuangjiang Lake Jiangdong Street
Mao Dian Resettlement Housing Project
Jinhua, Zhejiang
Residential

杭州临平昌达路车辆段上盖开发
浙江杭州
规划
Development of Hangzhou Linping
Changda Road Depot Superstructure
Hangzhou, Zhejiang
Planning

武汉城建·融创义乌江山云起
浙江义乌
居住
Wuhan UC & Sunac River Palace
Yiwu, Zhejiang
Residential

万科东莞臻山悦
广东东莞
居住
Vanke Dongguan Joy Hill
Dongguan, Guangdong
Residential

瑞安上海西藏南路项目
上海
城市设计
Shui On Shanghai South Xizang
Road Project
Shanghai
Urban Design

宜春樟树市樟药老街项目
江西宜春
文旅
Yichun Zhangshu Old Zhangyao
Street Project
Yichun, Jiangxi
Cultural Tourism

港中旅珠海平沙镇温泉大道项目
广东珠海
居住
CTSHK Zhuhai Pingsha Town Wenquan
Avenue Project
Zhuhai, Guangdong
Residential

日照山海田园小镇
山东日照
规划
Rizhao Shanhai Garden Town
Rizhao, Shandong
Planning

西奥杭州兴中路厂房
浙江杭州
产业
XIO Hangzhou Xingzhong Road Plant
Hangzhou, Zhejiang
Industry

飞鸟剧场
江苏无锡
文化
Earth Valley Theater
Wuxi, Jiangsu
Culture

绿城杭州下沙瑞华路住宅
浙江杭州
居住
Greentown Hangzhou Xiasha Ruihua
Road Residence
Hangzhou, Zhejiang
Residential

义乌建投后宅街道安置房
浙江金华
居住
Yiwu Construction Investment Houzhai
Street Resettlement Housing
Jinhua, Zhejiang
Residential

珑头湾旅游度假小镇
浙江温州
文旅
Longtou Bay Tourist Resort Town
Wenzhou, Zhejiang
Cultural Tourism

恒力海口江东新区酒店
海南海口
酒店
Hengli Haikou Jiangdong New District
Hotel
Haikou, Hainan
Hospitality

绿城深圳桂语兰庭
广东深圳
居住/商业
Greentown Shenzhen The
Osmanthus Grace
Shenzhen, Guangdong
Residential/Commercial

华润义乌下车门悦府二期
浙江义乌
居住
CR Land Yiwu Park Lane Complex Phase II
Yiwu, Zhejiang
Residential

绿城新桂沁澜
浙江宁波
居住
Greentown New Osmanthus Garden
Ningbo, Zhejiang
Residential

恒力苏州汾秋路办公楼
江苏苏州
办公
Hengli Suzhou Fenqiu Road Office
Suzhou, Jiangsu
Office

中交北京东小口马连店住宅
北京
居住
CCCG Beijing Dongxiaokou
Maliandian Residence
Beijing
Residential

融创周庄太史淀国际文旅城
江苏苏州
文旅
Sunac Zhouzhuang Taishidian
International Cultural Tourism City
Suzhou, Jiangsu
Cultural Tourism

绿城宁波洪塘西路住宅
浙江宁波
居住
Greentown Ningbo West Hongtao Road
Residence
Ningbo, Zhejiang
Residential

融创咸阳宸光壹号
陕西咸阳
居住
Sunac Xianyang One Mansion
Xianyang, Shaanxi
Residential

绿城桂语旗峰
广东东莞
居住/教育
Greentown Osmanthus Grace
Dongguan, Guangdong
Residential/Education

台州数字科技园
浙江台州
办公
Taizhou Digital Technology Park
Taizhou, Zhejiang
Office

绿城济南王府庄地铁上盖开发
山东济南
TOD
Greentown Development of Jinan
Wangfuzhuang Superstructure
Jinan, Shandong
TOD

钱塘湾未来总部基地
浙江杭州
城市设计
Qiantang Bay Future Headquarters Base
Hangzhou, Zhejiang
Urban Design

义乌建投华溪村游客中心
浙江金华
商业
Yiwu Construction Investment Huaxi
Village Tourist Center
Jinhua, Zhejiang
Commercial

蓝城承德金山岭项目
河北承德
居住
Bluetown Chengde
Jinshanling Residence
Chengde, Hebei
Residential

仁恒苏州仓街酒店概念设计
江苏苏州
酒店
Yanlord Suzhou Cang Street Hotel
Concept Design
Suzhou, Jiangsu
Hospitality

保利宜春明月川
江西宜春
居住
Poly Yichun Ming Yue Chuan
Yichun, Jiangxi
Residential

融创重庆云翠
重庆
商业
Sunac Chongqing Yuncui Mansion
Chongqing
Commercial

滨江栖江揽月轩
浙江杭州
居住
Binjiang Neo Mansion
Hangzhou, Zhejiang
Residential

雅达茗岭·窑湖小镇
江苏无锡
文旅
Yada Mingling Yaohu Town
Wuxi, Jiangsu
Cultural Tourism

绿城温州瓯江口雁鸣路项目
浙江温州
规划
Greentown Wenzhou Oujiangkou
Yanming Road Planning
Wenzhou, Zhejiang
Planning

绿城杭州燕语海棠轩
浙江杭州
居住/商业
Greentown Hangzhou Begonia
Hangzhou, Zhejiang
Residential/Commercial

恒德乡根台州黄岩小镇
浙江台州
规划
Hengde Xianggen Taizhou
Huangyan Town
Taizhou, Zhejiang
Planning

杭州传化科技城
浙江杭州
办公
Transfar Hangzhou Science and
Technology City
Hangzhou, Zhejiang
Office

中海常州翠语江南
江苏扬州
居住
China Overseas Changzhou Jiangnan
Courtyard
Yangzhou, Jiangsu
Residential

绿城大连海上明月
辽宁大连
居住
Greentown Dalian Bright Moon
Over the Sea
Dalian, Liaoning
Residential

既下山大同
山西大同
酒店
SUNYATA Hotel Datong
Datong, Shanxi
Hospitality

重庆融创湾BG地块
重庆
规划
Chongqing Sunac Bay Plot BG Project
Chongqing
Planning

上海花海芳香小镇木守酒店
上海
酒店
Shanghai Aroma Town Muh Shoou Hotel
Shanghai
Hospitality

绿城济南春风心语
山东济南
居住
Greentown Jinan Spring Mansion
Jinan, Shandong
Residential

三替职业技能培训中心
浙江杭州
办公
Santi Vocational Skills Training Center
Hangzhou, Zhejiang
Office

义乌建投北苑街道建设二村安置房
浙江金华
居住
Yiwu Construction Investment Beiyuan Street Construction Village II Resettlement Housing
Jinhua, Zhejiang
Residential

雅戈尔宁波集古路项目
浙江宁波
居住
Youngor Ningbo Jigu Road Residence
Ningbo, Zhejiang
Residential

仁恒江阴棠颂
江苏江阴
居住
Yanlord Jiangyin Tang Song
Jiangyin, Jiangsu
Residential

绿管宁波长乐未来社区
浙江宁波
居住
Greentown Construction Management Ningbo Changle Future Community
Ningbo, Zhejiang
Residential

交控绿城杭州月咏新辰轩
浙江杭州
居住
Taffic Control Technology & Greentown Hangzhou Yue Yong Xin Chen Xuan
Hangzhou, Zhejiang
Residential

滨汇杭州名望云筑
浙江杭州
居住
Binjiang & Golden Real Estate Hangzhou In One Mansion
Hangzhou, Zhejiang
Residential

东莞华润置地中心·悦府
广东东莞
混合开发
Dongguan CR Land Center · Park Lane Complex
Dongguan, Guangdong
Mixed-use

仁恒江阴棠颂
江苏江阴
居住
Yanlord Jiangyin Tang Song
Jiangyin, Jiangsu
Residential

绿城招商金华春熙明月
浙江金华
居住
Greentown & China Merchants Spring Scenery
Jinhua, Zhejiang
Residential

大拾花湾无锡太湖商墅
江苏无锡
规划
Danianhua Bay Wuxi Lake Tai Commercial Villa
Wuxi, Jiangsu
Planning

绿城哈尔滨诚园
黑龙江哈尔滨
居住
Greentown Harbin Sincere Garden
Harbin, Heilongjiang
Residential

融创苏高新常州樾澜庭
江苏常州
居住
Sunac & SND Group Changzhou China Mansion
Changzhou, Jiangsu
Residential

拾文投宜兴大拾花湾住宅
江苏宜兴
居住
Yixing Danianhua Bay Residence
Yixing, Jiangsu
Residential

浦发上海杨思路住宅
上海
居住
SPD Shanghai Yangsi Road Residence
Shanghai
Residential

明月松间朱泾酒店
上海
酒店
Yakamoz Hotel Zhujing
Shanghai
Hospitality

绿城台州湖境和庐
浙江台州
居住
Greentown Taizhou Oriental Villa
Taizhou, Zhejiang
Residential

绿城德润昆明柳岸晓风
云南昆明
居住
Greentown & Derun Kunming The Willow Shores
Kunming, Yunnan
Residential

绿城上海金山新城住宅
上海
居住
Greentown Shanghai Jinshan New City Residence
Shanghai
Residential

绿城南通市港闸区五水住宅
江苏南通
居住
Greentown Nantong Gangzha District Wushui Residence
Nantong, Jiangsu
Residential

弘安里
上海
城市更新 / 居住
Hong'anli
Shanghai
Urban Renewal/Residential

泰康之家·浙园
浙江杭州
康养
Taikang Community Zhe Garden
Hangzhou, Zhejiang
Health & Wellness

绿城北京万泉寺地块住宅
北京
规划
Greentown Beijing Wanquan Temple Lot Residence
Beijing
Planning

江南布衣萧山区厂房改造
浙江杭州
改造
JNBY Xiaoshan Plant Renovation
Hangzhou, Zhejiang
Renovation

浦发新杨思地块城市展示中心
上海
商业
SPD Bank New Yangsi Plot Urban Exhibition Center
Shanghai
Commercial

杭州灵隐寺佛学院二期
浙江杭州
文化
Hangzhou Lingyin Temple Buddhist College Phase II
Hangzhou, Zhejiang
Culture

南京江北新金融中心二期
江苏南京
城市设计
Nanjing Jiangbei New District Financial Center II
Nanjing, Jiangsu
Urban Design

绿城天津水西雲庐
天津
居住
Greentown Tianjin Yun Park
Tianjin
Residential

蓝城苏州春风湖滨
江苏苏州
居住
Bluetown Suzhou Lakeside Mansion
Suzhou, Jiangsu
Residential

绿城天津水西公园 E 地块
天津
居住
Greentown Tianjin Shuixi Park Plot E
Tianjin
Residential

绿城上海余庆里项目
上海
城市设计
Greentown Shanghai Yuqingli Project
Shanghai
Urban Design

珠海横琴天湖酒店
广东珠海
酒店
Zhuhai Hengqin Tianhu Hotel
Zhuhai, Guangdong
Hospitality

大拈花湾大有秋
江苏宜兴
文旅
Danianhua Bay Da You Qiu
Yixing, Jiangsu
Cultural Tourism

西奥上海松江厂房
上海
产业
XIO Shanghai Songjiang Plant
Shanghai
Industrial

金桥金港·星海湾
上海
居住
Golden Bridge The Ocean One
Shanghai
Residential

峨眉南山国际康养度假项目
四川乐山
酒店
Emei Nanshan International Wellness Resort Project
Leshan, Sichuan
Hospitality

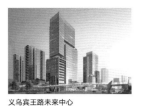

义乌宾王路未来中心
浙江义乌
混合开发
Yiwu Binwang Road Future Community
Yiwu, Zhejiang
Mixed-use

义乌商贸城集团荷塘名邸
浙江义乌
居住
Yiwu Trade City Group He Tang Ming Di
Yiwu, Zhejiang
Residential

融创咸阳高新区高科三路项目
陕西咸阳
居住
Sunac Xianyang High-tech Zone No.3 Gaoke Road Residence
Xianyang, Shaanxi
Residential

奥正朗园
山东临沂
居住
Auzen Lang Land
Linyi, Shandong
Residential

陆家嘴上海东站核心区住宅
上海
居住
Lujiazui Group Shanghai East Station Core Area Residence
Shanghai
Residential

中国电建地产中原华曦府
河南郑州
居住
Powerchina Real Estate Royal Mansion
Zhengzhou, Henan
Residential

义乌商贸城集团宝港路住宅
浙江义乌
居住
Yiwu Trade City Group Baogang Road Residence
Yiwu, Zhejiang
Residential

雅居乐中山翠亨新区 G42 地块酒店
广东中山
酒店
Agile Zhongshan Cuiheng New District Plot G42 Hotel
Zhongshan, Guangdong
Hospitality

华润郑州北龙湖瑞府
河南郑州
居住
CR Land Zhengzhou Majestic Mansion
Zhengzhou, Henan
Residential

蓝城杭州春风里
浙江海宁
居住/办公/商业
Bluetown Hangzhou Spring Link
Haining, Zhejiang
Residential/Office/Commercial

宜兴陶都科技新城经济产业园
江苏宜兴
产业/规划
Economic Industrial Park of Yixing Taodu Special District of Technology and Science
Yixing, Jiangsu
Industry/Planning

清河坊社区共富驿站
浙江杭州
混合开发
Qinghefang Community Center
Hangzhou, Zhejiang
Mixed-use

宸嘉 100 超高层公寓
湖北武汉
居住
Cityscape 100 Super-tall Residence
Wuhan, Hubei
Residential

中铁青岩健康小镇
贵州贵阳
康养
CREC Qingyan Health Town
Guiyang, Guizhou
Health & Wellness

翠著邱山里
浙江杭州
TOD
Hills Time
Hangzhou, Zhejiang
TOD

绿城前滩百合园
上海
居住
Greentown Jardin des Lys
Shanghai
Residential

杭州上城区城投钱江新城二期办公园区
浙江杭州
办公
Hangzhou Shangcheng Cheng Tou Qianjiang New City Phase II Office Park
Hangzhou, Zhejiang
Office

绿城北京西山云庐
北京
居住
Greentown Beijing Trees Villa
Beijing
Residential

上海新港国际中心
上海
居住/办公
Shanghai Xingang International Center
Shanghai
Residential/Office

华润绿城杭樾润府
浙江杭州
居住
CR Land & Greentown Hangyue Runfu
Hangzhou, Zhejiang
Residential

杭甬高速余姚服务区
浙江宁波
交通/规划
Yuyao Service Area of G92 Expressway
Ningbo, Zhejiang
Transpotation/Planning

2022

瑞安武汉天地
湖北武汉
混合开发
Shui On Land Wuhan Tiandi
Wuhan, Hubei
Mixed-use

仁恒海口美视高尔夫居住项目
海南海口
居住
Yanlord Haikou Meishi International Golf Club Residential Project
Haikou, Hainan
Residential

宁波杭州湾新区南洋小城安置房
浙江杭州
居住
Ningbo Hangzhou Bay New District Nanyang Town Resettlement Housing
Hangzhou, Zhejiang
Residential

宁波慈溪白沙路街道酒店
浙江宁波
酒店
Ningbo Cixi Baisha Road Sub-district Hotel
Ningbo, Zhejiang
Hospitality

华润杭曜置地中心
浙江杭州
混合开发
CR Land Hangzhou Center Crossing
Hangzhou, Zhejiang
Mixed-use

滦州竟鼎赛车场配套项目
河北唐山
酒店 / 居住
Luanzhou Jingding Automobile
Circuit Amenities
Tangshan, Hebei
Hospitality/Residential

腾讯上海华东总部
上海
办公 / 规划
Tencent Shanghai East China
Headquarters Project
Shanghai
Office/Planning

石梅自然体验馆
海南三亚
商业
Shimei Nature Experience Hall
Sanya, Hainan
Commercial

滨江杭州良渚新城公寓
浙江杭州
居住
Binjiang Hangzhou Liangzhu
New City Apartment
Hangzhou, Zhejiang
Residential

大拈花湾无锡水上拈花湾
江苏无锡
文旅
Danianhua Bay Wuxi Waterfront
Nianhua Bay
Wuxi, Jiangsu
Cultural Tourism

前湾控股宁波杭州湾新区南慈路居住宅
浙江宁波
居住
Qianwan Holdings Ningbo Hangzhou
Bay New District Nanci Road Residence
Ningbo, Zhejiang
Residential

总泉上海贝尚湾三期
上海
居住
Zongquan Shanghai Beishang Bay
Phase III
Shanghai
Residential

众安·岚荷芸府
浙江杭州
居住
Zhong'an Lan He Yun Fu
Hangzhou, Zhejiang
Residential

绿城春知海棠苑
浙江杭州
居住
Greentown Begonia
Hangzhou, Zhejiang
Residential

理想临平ESR地块
浙江杭州
规划
IDEAL Linping ESR Plot
Hangzhou, Zhejiang
Planning

仁恒济南港沟B-5/B-7地块住宅
山东济南
居住
Yanlord Jinan Ganggou
B-5/B-7 Plot Residence
Jinan, Shandong
Residential

前湾控股宁波杭州湾新区海洋一路项目
浙江宁波
混合开发
Qianwan Holdings Ningbo Hangzhou Bay
New District Haiyang No.1 Road Project
Ningbo, Zhejiang
Mixed-use

绿城宁波杭州湾新区芦汀路住宅
浙江宁波
居住
Greentown Ningbo Hangzhou Bay New
District Luting Road Residence
Ningbo, Zhejiang
Residential

鹏瑞云南大理健康养生创意园
云南大理
规划
Parkland Yunnan Dali Health
and Wellness Creative Park Planning
Concept Design
Dali, Yunnan
Planning

拈文投无锡拈花湾惠山古镇三期
江苏无锡
文旅
Wuxi Nianhua Bay Huishan Ancient
Town Phase III
Wuxi, Jiangsu
Cultural Tourism

滨江揽奥望座
浙江杭州
居住
Binjiang Vast Mansion
Hangzhou, Zhejiang
Residential

杭州龙兴广场立面改造
浙江杭州
改造
Hangzhou Longxing Square Facade
Renovation
Hangzhou, Zhejiang
Renovation

居然之家鄂州公园及商办项目
湖北鄂州
规划
Easyhome Ezhou Park and Commercial
Project
Ezhou, Hubei
Planning

华润广州白鹅潭悦府
广东广州
居住
CR Land Guangzhou Baietan CBD Project
Guangzhou, Guangdong
Residential

万科东莞臻湾汇
广东东莞
居住
Vanke Dongguan Elegant Mansion
Dongguan, Guangdong
Residential

滨江杭州北部新城项目
浙江杭州
居住
Binjiang Hangzhou North
New City Project
Hangzhou, Zhejiang
Residential

雅戈尔明湖懿秋
浙江宁波
居住
Youngor Wonder Lake
Ningbo, Zhejiang
Residential

滨江杭州富阳区富春街道住宅
浙江杭州
居住
Binjiang Hangzhou Fuyang Fuchun
Subdistrict Residence
Hangzhou, Zhejiang
Residential

西安高新独角兽企业创领中心
陕西西安
混合开发
Xi'an High-tech Unicorn Enterprise
Innovation Center
Xi'an, Shaanxi
Mixed-use

耀达临海市六角井未来社区
浙江临海
城市设计
Yaoda Linhai Liujiaojing Future
Community
Linhai, Zhejiang
Urban Design

绿城上海市黄浦区城市更新
上海
城市更新
Greentown Shanghai Huangpu
District Urban Renewal Project
Shanghai
Urban Renewal

中交映湖九里
浙江温州
居住
CCCG Super Center
Wenzhou, Zhejiang
Residential

华润北京悦府
北京
居住
CR Land Sino Palace
Beijing
Residential

溧阳金东方康养城
江苏常州
康养
Liyang Jindongfang Wellness City
Changzhou, Jiangsu
Health & Wellness

南昌九龙湖A08/A09/C03地块
江西南昌
规划
Nanchang Jiulong Lake A08/A09/C03 Plot
Nanchang, Jiangxi
Planning

陵水洲际酒店
海南陵水
酒店
InterContinental Lingshui
Lingshui, Hainan
Hospitality

华润成都锦江悦府
四川成都
居住
CR Land Chengdu Jinjiang Yuefu
Chengdu, Sichuan
Residential

农发杭州萧山新街里项目
浙江杭州
规划
ZJAD Group Hangzhou Xiaoshan Xinjieli Project
Hangzhou, Zhejiang
Planning

源创西安大明宫八宗地块项目
陕西西安
混合开发
Source Group Xi'an Daming Palace Eight Plots Project
Xi'an, Shaanxi
Mixed-use

陆家嘴锦绣观澜
上海
居住
Lujiazui Group Splendid Mansion
Shanghai
Residential

仁恒无锡夹城里AD地块
江苏无锡
居住
Yanlord Wuxi Jiachengli Plot AD Residence
Wuxi, Jiangsu
Residential

华发珠海四季云山
广东珠海
居住
Huafa Zhuhai Sky Hill Mansion
Zhuhai, Guangdong
Residential

杭州康基奇璞医疗器械有限公司研发中心
浙江杭州
办公
Hangzhou Kangji CHIP Medical Instrument Co., Ltd. R&D Center
Hangzhou, Zhejiang
Office

绿城银川阅海湾壹号
宁夏银川
居住
Greentown Yinchuan Honor
Yinchuan, Ningxia
Residential

象山城投·绿城N30°梦想城
浙江宁波
居住
Xiang Shan Municipal Construction Investment & Greentown N30° Imagine City
Ningbo, Zhejiang
Residential

宸嘉上海长风9号地块住宅
上海
居住
Cityscape Shanghai Changfeng Plot 9 Residence
Shanghai
Residential

绿城杭州馥香园
浙江杭州
居住
Greentown Hangzhou The Blooming Park
Hangzhou, Zhejiang
Residential

灵峰智慧谷
浙江湖州
产业/办公
Lingfeng Smart Green Valley
Huzhou, Zhejiang
Industrial/Office

绿城桂香园
浙江杭州
居住
Greentown TAOLI
Hangzhou, Zhejiang
Residential

杭州Club-Med度假酒店
浙江杭州
酒店
Club Med Hangzhou
Hangzhou, Zhejiang
Hospitality

上海浦东新区前滩南滨江14单元二期
上海
居住
Shanghai Pudong New Area Qiantan South Riverside Unit 14 Phase II Residence
Shanghai
Residential

徐汇城投上海天平街道小襄阳住宅
上海
城市更新/规划
Xuhui Urban Investment Shanghai Tianping Subdistrict Xiaoxiangyang Residence
Shanghai
Urban Renewal/Planning

中海上海闵行古美地块住宅
上海
居住
Zhonghai Shanghai Minhang Gumei Plot Residence
Shanghai
Residential

绿城宁波鄞州文启路住宅
浙江宁波
居住
Greentown Ningbo Yinzhou Wenqi Road Residence
Ningbo, Zhejiang
Residential

泰康之家·浦园
上海
医养
Taikang Community Pu Garden
Shanghai
Health & Wellness

博多杭州森谷产业园
浙江杭州
产业/办公
Boduo Hangzhou Sengu Industrial Park
Hangzhou, Zhejiang
Industrial/Office

上海徐汇区漕宝路亿滋地块
上海
居住
Shanghai Xuhui District Caobao Road Project
Shanghai
Residential

杭州市上城区南星桥馒头山街区更新
浙江杭州
城市更新/规划
Hangzhou Shangcheng Nanxingqiao Mantoushan Block Renewal
Hangzhou, Zhejiang
Urban Renewal/Planning

无锡太湖锦绣园二期
江苏无锡
居住
Wuxi Lake Vista Phase II
Wuxi, Jiangsu
Residential

安吉灵峰国际养心小镇
浙江湖州
文旅
Anji Lingfeng international heart town
Huzhou, Zhejiang
Cultural Tourism

耀达临海珑玺府
浙江台州
居住
Yaoda Linhai Longxifu
Taizhou, Zhejiang
Residential

宁波溪口文旅项目
浙江宁波
商业
Ningbo Xikou Cultural Tourism Project
Ningbo, Zhejiang
Commercial

宁波杭州湾新区体育公园5#地块
浙江宁波
混合开发
Ningbo Hangzhou Bay New District Sports Park Plot 5
Ningbo, Zhejiang
Mixed-use

华润三亚海棠湾住宅
海南三亚
居住
CR Land Sanya Haitang Bay Residential Project
Sanya, Hainan
Residential

四川德格格萨尔王文化中心
四川甘孜
文化
Sichuan Dege King Gesar Cultural Center
Ganzi, Sichuan
Culture

郑州郑东新区北龙湖住宅
河南郑州
居住
Zhengzhou Zhengdong New District Beilong Lake Residence
Zhengzhou, Henan
Residential

日照利伟山海天路项目
山东日照
居住
Rizhao Liwei Shanghaitian Road Project
Rizhao, Shandong
Residential

陶都科创中心
江苏宜兴
办公
Taodu Scientific and Technological Innovation Center
Yixing, Jiangsu
Office

仁恒南京江宁住宅
江苏南京
居住
Yanlord Nanjing Jiangning Residence
Nanjing, Jiangsu
Residential

华和杭州北部新城LZ10单元超高层办公楼
浙江杭州
办公
Wah Wo Hangzhou North New Town LZ10 Unit Super High-rise Office Building
Hangzhou, Zhejiang
Office

绿城西安未央WY9-13-385地块项目
陕西西安
居住
Greentown Xi'an Weiyang Plot Y9-13-385 Project
Xi'an, Shaanxi
Residential

无锡夹城里幼儿园
江苏无锡
教育
Wuxi Jiachengli Kindergarten
Wuxi, Jiangsu
Education

绿城汀岸辰风里
浙江杭州
居住
Greentown Wave Wind
Hangzhou, Zhejiang
Residential

中能建上海金山项目
上海
居住
CEEC Shanghai Jinshan Project
Shanghai
Residential

温州平阳坡南街项目
浙江温州
城市更新 / 酒店
Wenzhou Pingyangpo South Street project
Wenzhou, Zhejiang
Urban Renewal/Hospitality

华润长隆万博悦府
广东广州
居住
CR land Park Lane Complex
Guangzhou, Guangdong
Residential

恒力苏州市盛泽镇住宅
江苏苏州
居住
Hengli Suzhou Shengze Town Residence
Suzhou, Jiangsu
Residential

义乌市宾王路双创中心
浙江义乌
混合开发
Yiwu Binwang Road Innovation and Entrepreneurship Center
Yiwu, Zhejiang
Mixed-use

2023

绿城义乌福田街道稠州北路西北侧地块
浙江义乌
居住
Greentown Yiwu Futian Neighborhood Chouzhou North Road Northwest Plot Project
Yiwu, Zhejiang
Residential

南京六合雄州新城二期住宅
江苏南京
居住
Nanjing Liuhe Xiongzhou New City Phase II Residence
Nanjing, Jiangsu
Residential

杭州三里亭幼儿园
浙江杭州
教育
Hangzhou Sanliting Kindergarten
Hangzhou, Zhejiang
Education

蓝城赣州峰韵龙井小镇酒店
江西赣州
酒店
Bluetown Ganzhou Fengyun LongjinTown Hotel
Ganzhou, Jiangxi
Hospitality

绿城临平新城核心区36号地块
浙江杭州
规划
Greentown Linping New City Core Area Lot 36 Planning
Hangzhou, Zhejiang
Planning

中吴江南春
江苏常州
居住
Zhongwu Jiannan Spring
Changzhou, Jiangsu
Residential

浙江禄筌胜实业有限公司办公产业园
浙江杭州
产业 / 办公
Zhejiang Luquan Sheng Industrial Co., Ltd. Industrial park
Hangzhou, Zhejiang
Industrial/Office

华友控股广西玉林博白县白平产业园
广西玉林
产业
Guangxi Yulin Huayou Holding Baiping Industrial Park, Bobai County
Yulin, Guangxi
Industrial

溧阳天目湖动物王国酒店
江苏溧阳
酒店
Liyang Tianmu Lake Animal Kingdom Hotel
Liyang, Jiangsu
Hospitality

华润苏州中新大道南住宅
江苏苏州
居住
CR Suzhou Zhongxin Avenue South Residence
Suzhou, Jiangsu
Residential

华润华发翡云润府
浙江杭州
居住
CR Land & Huafa Only Island Over Qianjiang
Hangzhou, Zhejiang
Residential

海底捞成都A1地块
四川成都
酒店
Haidilao Chengdu Lot A1 Project
Chengdu, Sichuan
Hospitality

滨江越秀中豪杭州翠宸里
浙江杭州
居住
Binjiang Yuexiu Zonhow Jadeite
Hangzhou, Zhejiang
Residential

阿那亚隆化汤泉酒店
河北承德
酒店
Anaya Longhua Hot Spring Hotel
Chengde, Hebei
Hospitality

杭州钱塘区江海商务中心
浙江杭州
办公
Hangzhou Qiantang District Jianghai Business Center
Hangzhou, Zhejiang
Office

郑州金桥北龙湖117商业项目
河南郑州
混合开发
Zhengzhou Jinqiao North Longhu 117 Commercial Project
Zhengzhou, Henan
Mixed-use

建发杭州东方示范区项目
浙江杭州
商业
C&D Hangzhou East Demonstration Zone Project
Hangzhou, Zhejiang
Commercial

宝业绍兴柯桥鉴水路住宅
浙江绍兴
居住
Baoye Shaoxing Keqiao Jianshui Road Residence
Shaoxing, Zhejiang
Residential

科维常州横山桥三山田园项目一期
江苏常山
规划
KEWEI Changzhou Hengshan Bridge Sanshan Farm Project phase I
Changshan, Jiangsu
Planning

保利半岛1号
海南万宁
居住
Poly Peninsula No.1
Wanning, Hainan
Residential

济高控股济南历城区世纪大道住宅
山东济南
居住
Jinan Hi-Tech Holding Group Jinan Licheng Century Avenue Residence
Jinan, Shandong
Residential

济宁能源嘉和康泰城颐园
山东济宁
居住
Jining Energy Jiahe Kangtai City Yi Garden
Jining, Shandong
Residential

蓝城济南中央商务区西片区A5/A10地块
山东济南
居住
Bluetown West of Jinan Business District Plot A5/A10
Jinan, Shandong
Residential

华润温州鹿城区东游路住宅
浙江温州
居住
CR Wenzhou Lucheng District East You Road Residence
Wenzhou, Zhejiang
Residential

吴江盛家库再生规划项目
江苏吴江
城市更新
Wujiang Shengjiashe Regeneration Planning Project
Wujiang, Jiangsu
Urban Renewal

中海月珑云岚
浙江杭州
居住
China Overseas Chea Sinensis
Hangzhou, Zhejiang
Residential

义乌星城办公楼
浙江金华
办公
Yiwu Xingcheng Office Building
Jinhua, Zhejiang
Office

理想临平老城区 LP0605-04 地块住宅
浙江杭州
居住
IDEA Linping Old Town Plot LP0605-04 Residence
Hangzhou, Zhejiang
Residential

大横琴湖心新城
广东珠海
城市设计
Da Heng Qin Huxin New City
Zhuhai, Guangdong
Urban Design

蓝城江苏吴江九院地块项目
江苏吴江
居住
Bluetown Jiangsu Wujiang Jiuyuan Plot Project
Wujiang, Jiangsu
Residential

东苑上海市松江区桃园路三号地块项目
上海
居住
Dongyuan Shanghai Songjiang District Taoyuan Road Plot No.3 Project
Shanghai
Residential

绿城台州仙居神仙居景区文旅住宅
浙江台州
规划
Greentown Taizhou Xianju Shenxianju Scenic Area Cultural Tourism Residence
Taizhou, Zhejiang
Planning

万科厦门五缘湾商业项目
福建厦门
商业
Vanke Xiamen Wuyuan Bay Commercial Project
Xiamen, Fujian
Commercial

西子杭州三堡 JG1203-73 地块项目
浙江杭州
混合开发
Xizi Hangzhou Sanbao JG1203-73 Plot Project
Hangzhou, Zhejiang
Mixed-Use

华润成都陆肖 31 亩住宅项目
四川成都
居住
CR Land Chengdu Luxiao 31mu Residential Project
Chengdu, Sichuan
Residential

临安天目山自然生态研学城一期
浙江杭州
混合开发
Lin'an Tianmu Mountain Natural Ecology Research City Phase I
Hangzhou, Zhejiang
Mixed-Use

华润长沙雨润瑞府
湖南长沙
居住
CR Land Crest Residence
Changsha, Hunan
Residential

开投蓝城台州黄岩区南城街道住宅
台州黄岩
居住
Kaitou Bluetown Taizhou Huangyan District Nancheng Street Residence
Huangyan, Taizhou
Residential

绿城长沙滨湖新城项目
湖南长沙
城市设计
Greentown Changsha Binhu New City Project
Changsha, Hunan
Urban Design

蓝城杭州天都城商业项目
浙江杭州
混合开发
Bluetown Hangzhou Tiandu City Commercial Project
Hangzhou, Zhejiang
Mixed-Use

伟星芜湖翡翠天元
安徽芜湖
居住
Weixing Wuhu Fei Cui Tian Yuan
Wuhu, Anhui
Residential

瑞玺瑞安市东山西单元 03-34 地块住宅
浙江温州
居住
Ruixi Rui'an Dongshan Western Unit 03-34 Plot Residence
Wenzhou, Zhejiang
Residential

绿城玉海棠
浙江杭州
居住/办公
Greentown Yu Hai Tang
Hangzhou, Zhejiang
Residential/Office

中交郑州北龙湖住宅
河南郑州
居住
CCCG Zhengzhou Beilonghu Residence
Zhengzhou, Henan
Residential

绿城上海乔家路城市更新项目
上海
城市更新
Greentown Shanghai Qiaojia Road Urban Renewal Project
Shanghai
Urban Renewal

济高控股济南历下区恩翼帕瓦地块项目
山东济南
居住
Jinan Hi-Tech Holding Group Jinan Lixia District Enyi Pawa Plot Project
Jinan, Shandong
Residential

方远集团台州东城街道住宅
浙江台州
居住
Fangyuan Taizhou Dongcheng District Residence
Taizhou, Zhejiang
Residential

绿城杭州萧山区市北单元 XSCQ1309-R2-50 地块项目
浙江杭州
居住
Greentown Hangzhou Xiaoshan District Shibei Unit XSCQ1309-R2-50 Plot Project
Hangzhou, Zhejiang
Residential

德港乌鲁木齐科教生态小镇
新疆乌鲁木齐
规划
Degang Urumqi Science and Education Eco-town
Urumqi, Xinjiang
Planning

华润深圳润宏城
广东深圳
居住
CR Land Dream City
Shenzhen, Guangdong
Residential

大横琴珠海 15号/16号/16A号/16B号地块项目
广东珠海
居住
Da Heng Qin Zhuhai 15#/16#/16A#/16B# Plot Project
Zhuhai, Guangdong
Residential

杭州灵隐寺湖州莫干山天池禅寺
浙江湖州
文化
Hangzhou Lingyin Temple Huzhou Mogan Mountain Tianchi Zen Temple
Huzhou, Zhejiang
Culture

蓝城嵊州剡湖街道 RB2023-28 地块住宅
浙江嵊州
居住
Bluetown Shengzhou Suanhu Street RB2023-28 Plot Residence
Shengzhou, Zhejiang
Residential

绿城南京河西绿博园地块项目
浙江南京
居住
Greentown Nanjing the Green Expo Garden Plot Project
Nanjing, Zhejiang
Residential

融创杭州新塘住宅项目
浙江杭州
居住
Sunac Hangzhou Xintang Residential Project
Hangzhou, Zhejiang
Residential

图书在版编目（CIP）数据

大象设计：2018—2023：汉文、英文 / goa 大象设计编著.—上海：同济大学出版社，2023.10
ISBN 978-7-5765-0923-6

Ⅰ. ①大… Ⅱ. ①g… Ⅲ. ①建筑设计 - 作品集 - 中国 - 现代 Ⅳ. ①TU206

中国国家版本馆 CIP 数据核字 (2023) 第 181796 号

策划 & 编纂团队

总　策　划：卿州　程思
视觉统筹：陈欣彤　Maxime LAURENT
装帧设计：七月合作社
出版统筹：许静瑶
文字编辑：程思　许静瑶　蒋嘉菲　李凌　陈欣彤　方舟　李夏
制　　图：佘纤言　麦思琪　聂微雨　周澳　邱锡伶　徐佳惠　周伊杰　陈薇伊　何松　李政
翻　　译：佘纤言　张雨婷　张谱
责任编辑：徐希
责任校对：徐逢齐
英文校对：Stephen P. Davis

大象设计 2018—2023
goa 大象设计　编著

出版发行：同济大学出版社（地址：上海市四平路 1239 号　邮编：200092　电话：021-65985622）
经　　销：全国各地新华书店、建筑书店、网络书店
开　　本：889mm×1194mm　1/12
印　　张：25.33
字　　数：790 000
版　　次：2023 年 10 月第 1 版
印　　次：2023 年 10 月第 1 次印刷
书　　号：ISBN 978-7-5765-0923-6
定　　价：369.00 元

版权所有　侵权必究
印装问题　负责调换